Computers and the Psychosocial Work Environment

Gunilla Bradley
Institute of International Education
University of Stockholm

Translated by Struan Robertson

Taylor & Francis
London · New York · Philadelphia
1989

UK Taylor & Francis Ltd, 4 John St, London WC1N 2ET

USA Taylor & Francis Inc., 242 Cherry St, Philadelphia, PA 19106-1906

British Library Cataloguing in Publication Data

Bradley, Gunilla
Computers and the psychosocial work environment.
1. Quality of life. Effects of technological innovation
I. Title II. Psykosocial arbetsmiljö och datorer. *English*
303.4'83

ISBN 0–85066–455–1
ISBN 0–85066–457–8 Pbk

Library of Congress Cataloging-in-Publication Data

Bradley, Gunilla
Computers and the psychosocial work environment.
Bibliography: p.
Includes index.
1. Labor supply—Effect of technological innovations on. 2. Computers—Social aspects. I. Title.
HD6331.B69 1988 331.25 88–2298
ISBN 0–85066–455–1
ISBN 0–85066–457–8 (pbk.)

Cover design by Jordan and Jordan, Fareham, Hampshire
Typeset in Bembo 11/12 pt by Photo·Graphics, Honiton, Devon
Printed in Great Britain by Taylor & Francis (Printers) Ltd, Basingstoke, Hampshire

Contents

Preface

Computer technology and its applications have a considerable impact on the structure of both our working lives and our social lives. Computer technology is in a period of accelerating development and is making inroads into more and more areas of our society. Both hardware (the equipment itself) and software (the programs used in computers) bring about changes in our lives, affecting the labour market as a whole, the operations of individuals, companies and the conditions under which we work and live.

Each one of us can, on the basis of our individual circumstances and conditions, become actively involved in this process of change. We are still in a period of upheaval. Many of us have still not adjusted to the computer society, and in this situation we are able to pose questions and point to desirable changes. The closer we get to a fully computerized society the harder it becomes to have any influence on events. When we are in the middle of a sweeping change, our awareness of the process of change is often diminished. Changes often occur in small steps which, taken together, constitute a 'leap' forward.

In this book, I summarize the results and experience gained from 'The Effects of Rationalization on the Work Environment and Working Conditions in Some Salaried Employee Groups', known as the RAM Project, which I headed at the Department of Sociology, University of Stockholm, from 1975 to 1985. The findings of each stage of the RAM project are published in a series of reports, RAM reports 1–12, and in various other forms. I have presented some of the preliminary findings in the books, *Occupational Roles and the Quality of Life* (1979) and *Computer Technology, Working Life and Communication* (1977). The experiences gained from research at an insurance company are to be found in *The Work Environment and Display Terminals* (Bradley, 1981). These last two books have been unavailable for the last two years. There has been a growing interest in the experience and the findings from this project and for that reason I have attempted to summarize those findings in this book.

The aim of the RAM Project was to make a more detailed analysis of the psychosocial work environment and its link with administrative rationalization, primarily with changes brought about by advances in electronic data processing (EDP) technology.

At the beginning of the 1970s, when I carried out a problem inventory of the work environments of salaried employees, rationalization—mainly in the form of computerization—was mentioned as one of the dominant changes. I then produced a research plan and contacted the then newly-established Swedish Work Environment Fund, which financed my research.

The first RAM Project is referrd to in the book as RAM:1. There were two sub-projects within the RAM:1 Project, which studied two different organizations:

Project I was based on studies carried out at a state-owned corporation, the Postal Giro Centre (organization 1). The results of this work are reproduced mainly in Chapter 4.

Project II was based on studies at the Trygg-Hansa insurance company (organization 2). The results of this work are to be found mainly in Chapters 5 and 8.

In technical terms, two main types of system were examined, each of which has a different effect on the structure of the work environment. They may be related to different phases in 'the history of computer technology', and are:

- the so-called batch-type system;
- an online type of system using display terminals.

The problem areas that were highlighted cover a broad area and are linked to the concepts in the theoretical models in Chapter 3. In brief, they are as follows:

- the organizational structure;
- work content;
- influence at work;
- communication patterns;
- the intermediate position/buffer role;
- management/the role of manager;
- working hours;
- wage forms.

In the final phase of the RAM Project, I produced a new major programme for a research project on computer technology and changes in working life. This research is currently in progress, and is referred to in this book as the RAM:2 Project. This project focuses on studies and analyses of microprocessor systems—a third phase in the development of this technology. It also concentrates on three areas which, on the basis of the earlier research, are considered to constitute the key problems related to the development of computer technology in working life. In addition to work content, these analyses include:

- the structure of communications at work;
- professional roles—management/colleague;
- working life—the conditions of life 'outside work'.

Chapter 6 presents a sub-study from the RAM:2 Project. This study was conducted at an electronics company and deals with the changing role of the secretary as a result of the introduction and use of a word processing system.

This book may be read in a number of ways. The first chapter discusses the opportunities and the risks inherent in computer technology. Here I have tried to place some of the results from the RAM Project in the context of the current level of knowledge about the development of computer technology and the current debate on this issue. Chapter 2 contains a brief review of the development of computer technology, the present position and future developments. Telecommunications technology is also discussed here, where the symbiosis of computer technology and teletechnics forms a powerful force behind changes in working life. Chapter 3 is aimed at readers who want more details of the background, the issues involved and the theoretical models and methods of the RAM Project. Chapters 4, 5 and 6 will be of greatest interest to people whose main concern is the results of the research conducted in these three companies. Chapters 4 and 5 begin with technical descriptions of the computer systems that were studied. At the end of Chapter 5 there are some follow-up interviews with key staff at the company.

Chapters 4 and 5 describe the quantitative measurement instrument used to obtain the empirical results (in particular, the evaluation of the terminal system). For more details of the content and the structure of this questionnaire and instructions on its use, see Bradley (1981) *The Work Environment and Display Terminals* (in Swedish), available from large libraries in Sweden. A number of indices were designed on the basis of this questionnaire, which were thought to be more reliable than simple questions because of their wider scope. The indices measure different problem areas, the psychosocial aspects of the work environment, the effects of the display terminal system and of other forms of computerization, democracy at the workplace, health factors and the weighting of different work environment factors. It is often important to evaluate the work environment when introducing new computer systems or when examining systems that are already in use, but desirable working conditions should also be defined and set out. Several employer-employee agreements now state that the introduction of a computer system must be followed by an assessment of its effects on conditions of work. This applies to major changes of systems already in operation as well as to the planning and introduction of new computer systems.

The research covered a broad spectrum of areas in the work environment (indices). The indices measure the areas in the work environment which proved to be relevant to the individual's well-being at work and which are important to consider when introducing computer technology at a workplace. One can select the areas of the work environment that are important to consider and that may have proved to be significant in a more qualitative type of preliminary study (see, for example, the RAM:1 report, which also presents a questionnaire that covered central organizational variables and computerization).

Chapter 7 highlights a key problem of computerization: stress and mental strain. In part, it repeats the results set out in Chapters 4 and 5, and can be read separately from the rest of the book. Chapter 8 discusses the significance of the psychosocial work environment in terms of the benefits we derive from our leisure time and family lives and in terms of the individual's health.

Chapter 9 deals with more general issues in the sphere of computerization and the work environment, and proposes action that relates more directly to the results of the RAM Project, mainly the work at Trygg-Hansa. There is also a description of the follow-up work that was carried out in stages at Trygg-Hansa. There is a similar section on the Postal Giro Centre at the end of Chapter 4.

In Chapter 10 I put forward some proposals for research in this area that I consider to be important, and offer some personal reflections on research work and the key questions for the future vis-à-vis our working lives and private lives.

In Appendix 1, I reproduce the main content of the various empirical measurements for the evaluation of the work environment when computer technology is introduced. Appendix 2 is a list of the reports and books originating from the RAM Project.Explicit references to this material have sometimes been omitted from the text to avoid making it difficult to read. Appendix 3 is a checklist on effects on the work environment that can be used in assessing computer systems. Appendix 4 provides a list of reports I have prepared indicating the priorities for research in the field of changes in working life. Appendix 5 on labour relations in Sweden provides details of the broader context in which this study was conducted.

The experience and results gained from the RAM Project have become widely-known. They have been used in various ways and at various levels in work on computerization and the work environment and have also given rise to more research in this field.

I should like to express my thanks to the Work Environment Fund for funding my research. My thanks to friends and colleagues for their stimulating cooperation, characterized by the fact that we constantly learned from one another. Thanks to Willy, Linda and Karin. Good friendships, good relationships with colleagues and the constant support of my family have been important to me in writing this book.

Gunilla Bradley
Stockholm, August 1988

1. The consequences of computerization: opportunities and risks

1. Employment and job structure

The rate of employment and the structure of the labour force in our society are emphasized as important areas in current discussions on computerization. They touch upon the fundamental human need for security—in the broader sense of the term. Two Swedish government commissions have investigated these factors.

As is the case for most areas when technology gains in influence through computerization, it affects employment in many ways; this, of course, has also been the purpose of rationalization. The main impact of the introduction of data processing technology thus includes the disappearance of some jobs and the non-appearence of others that would otherwise have emerged. But it should also be noted that various new jobs have been created as a direct result of this new technology.

In the area of office work, such developments have ranged from computerization of simple routines to that of entire functions, such as accounting, order management, invoicing. A major impact on the rate of employment is expected when this technology is increasingly adopted in connection with the growing use of text and word-processing systems. The present trend towards technically advanced and integrated communication and information systems will probably result in a fundamental change of the entire office world, affecting its employment structure and work organization, as well as its work environment in a broad sense.

Qualified areas of work are being affected, even in technological and economic fields such as cost estimates and design work, budgeting, production engineering and planning. Its directly measurable quantitative effects on the rate of employment have been relatively small to date. The larger companies have implemented data processing technology to the greatest extent, and they have met a reduced need for employees by transferring them or awaiting their so-called 'natural retirement' (which has not always been so natural). In addition, the projected goals of such rationalization will probably be reached only in a long-range perspective, when the new technology is entirely integrated into the organization. Thus, the future impact on employment will be even more striking. 'The role of the public sector' has also become one of the key issues in the Swedish debate in this field.

In discussing these matters, some economists have claimed that the economy at the national level is so complex that it is impossible to conclude with any certainty which changes in the sphere of employment are caused by the introduction of computers and which are caused by other factors. At this point, an attempt is made to judge 'the rationalization potential of computerization' and then to examine more closely how likely it is that this potential is being exploited. In the search for knowledge in this difficult field, it is natural to use not only broad statistical analyses of employment trends within different sectors and occupational groups on the labour market, but also analyses of the future scenario type.

Many people are of the opinion that computerization mainly poses a threat to employment, while others see the introduction of data processing technology as a means of stimulating employment. It is necessary to try to make a realistic assessment of its impact on employment and discuss at an early stage measures and strategies to handle this development, and carefully consider the implementation of new technology for purposes that are determined by the actual users.

Rationalizations by means of data processing and employment issues will become of central interest here, due to the character of the industrial investments being made. The forms of investments that dominate western industrial societies are those which lead to the rationalization of the production of goods and services. On the other hand, investments aimed at increasing production have not developed equally rapidly. Perhaps we should discuss the total automation of some branches of industry where work environment requirements are difficult to meet, as well as investments in an entirely new type of production which instead improves the environment.

Of central importance for the future labour market, both in terms of employment and of the labour force structure, is the rapidly accelerating use of microcomputers. It is especially difficult to estimate the impact of microcomputerization on employment, due to its inherent character of 'invisibility', which is linked to its 'smallness' and 'cheapness'. Local initiatives for the purchase of microcomputer-based equipment make an overview more difficult.

The problem area of computerization employment leads to a more general key question for the future—the question of allocation. This need not be seen as a form of resignation or pessimism, but rather as optimism. It all depends upon what we have to allocate and how we value it. International competition leads to rationalization, and computerization today is easily one of the most important aids in rationalization. Whether seen in a national or an international perspective, computerization entails a series of allocation questions:

- allocation of profit
  - between sectors within a country,
  - between industrialized countries,
  - between industrialized and third world countries.

- allocation of work (shortening of working hours, division of labour, etc.)
  - allocation of leisure time
  - allocation of compulsory citizen services (paid or unpaid)
  - allocation of production and reproduction.

What goals do we wish to set up for the labour market and employment?

- Do we want a two-tier labour force, divided into an A-level and B-level—one level of workers whose specialist skills make them difficult to replace when sick; and another level of workers with less knowledge, who can be easily replaced and who risk losing their jobs in the next phase of the automation process?
- Do we want to allocate existing jobs, shorten the working hours per day and, in the long run, enjoy the 'Brave, New Everyday'?
- Can we develop new products, goods and services within totally new areas, which will repair the destruction caused by industrialization? Can we create new goods and services in which the experiences and imagination of women will be utilized more than is now the case, and perhaps even relinquish some of the current production of goods and services?
- Shall we look at women's and men's professional work/activities from a life-cycle perspective and recognise certain occupationally active periods in life?
- Within which areas of working life and private life should we not have computer technology at all (taking into account ethics and considerations of quality)?
- Is a system of recurrent education an appropriate pedagogical principle for the future? Perhaps education should be seen as an end in itself instead of a means to an end?
- Should self-esteem be based on working life or private life (in the terminology in current use)?

- If computers are now an aid to rationalization, which of these rationalization philosophies ought to apply:
  - the strict division of labour, vertically and laterally?
  - the creation of work tasks considered to be personally enriching and which serve the individual's as well as the company's purposes, and then let machines take over the rest?

Isn't it about time we used rationalization ideologies which respect the total life situations of women and men?

2. Some reflections on the organization of work

The structure of an organization forms the framework for changes in individual job positions or in work tasks. The design of the organization is a crucial factor in the study of the relationship between data processing and work environment, and this same organizational design can also be seen as a direct part of the work environment. The organizational structure is defined as methods for dividing labour, the grounds for decision-making, organizational aids and so on.

Changes in the structure of the organization may be a direct objective of computerization. A company may perhaps want to expand some units/operations and to reduce others, to eliminate some levels via rationalizations, to steer and control others via data processing, to localize some departments or companies within the concern, or even to decentralize or centralize the operations of the company. It is also possible for employees to develop data processing systems that they need on a local level, and for them to cooperate with other sections of the operation. In this way, increased job autonomy can be developed on a local basis. However, changes in the organizational structure often become more indirect effects, unintentional effects, of the use of data processing. The organizational design already in existence before computerization is often strengthened.

Future use of telecommunication techniques in combination with microcomputer-based equipment of various types (for example, word processing systems, CAD/CAM systems, personal computers) will create new professional roles and patterns of communication between individuals and vocational groups. This will result indirectly in a 'new look' both for the formal organizational chart (a well-formed system with arrows and boxes) and for the organizational structure in the form of informal networks. Such developments should be planned carefully and be based on knowledge about people and organizations, and the interplay between them.

What has happened hitherto to organizational structure in connection with computerization? Three phases in the development of technology deserve comment. In the late 1950s and especially in the early and mid-1960s, the dominant philosophy proclaimed that data processing power should be centralized—it was the golden age of the mainframe computer. As a result, larger and faster computers were designed with relatively high memory capacities, and used more for purely administrative applications than for scientific or technical applications. Data processing departments began to proliferate.

During the latter part of the sixties, this trend was broken, as minicomputers appeared on the scene. They had limited memory and calculation capacities, but were fully capable of handling numerous technical applications and tasks. Moreover, they were inexpensive and easier to install and handle than the large all-purpose computers. Minicomputers increasingly resembled mainframe computers in performance, but were available at much lower prices.

Then, in a third phase of this development, the microcomputer was devised. The microcomputer became the major contribution of component technology to the needs of the 1970s for decentralized small computers which could be built directly into instruments, machines and other equipment.

Microprocessors, 'the heart' of the microcomputer, can be characterized as: highly complex, but easy to adapt to a wide variety of tasks; mass-produced, but adaptable to final applications by users who produce the necessary software themselves; and very inexpensive to purchase, but imparting a higher value and usefulness to the products in which they are placed.

A large research project (the RAM Project) was conducted from 1975 to 1985 at the University of Stockholm, financed by the Swedish Work Environment Fund. Its objective was to analyze the psychosocial work environment and to examine its relationship with changes which occur through administrative rationalizations, mainly those employing computer technology. The first company studied, the Postal Giro Service, is an example of an organization which was exclusively using mainframe computers and a so-called batch system. Below is a list of some changes in the design/structure of the organization that it was possible to attribute to the computer system:

- a decrease in the number of levels in the hierarchy, due to the disappearance of some positions;
- the closing down of a special key-punching centre in the organization;
- the expansion of the data processing department;
- the creation of a special department for correcting errors made during handling of data.

The RAM Project also observed the introduction and use of display terminals within an insurance company (Trygg-Hansa), where both mainframe and minicomputers were used. The following changes in the structure of the organization were observed:

- a tendency towards eliminating a number of levels in the organization; there had been too many supervisors on the intermediate level, a result of numerous mergers;
- enlargement of a data processing department, especially the commissioning of a large number of outside consultants;
- introduction of a special word processing department;
- even before terminalization, the company had democratized its structure, including the introduction of leadership groups, etc., which facilitated some of the changes; thus a more appropriate organizational structure had been established before the arrival of the technology;
- the simultaneous introduction of modern office design and terminalization meant the collision of these two different methods of rationalization; for example, terminalization placed special demands on office lighting, demands which were hard to reconcile with many office landscape layouts. This led to the placement of floor screens around the display terminals.

The following are specific implications of the use of microcomputers, the third phase in the development of technology:

- increased dependency on other countries, and thus national vulnerability;
- the enormous growth potential created by 'cheapness' and 'smallness';
- the invisible character of 'microcomputerization', as compared to 'mainframe computerization';
- major opportunities for decentralization;
- major opportunities for centralization, such as linkage with computer systems and databases;
- opportunities for local initiatives—other strategies for implementation of the technology.

The question of centralization and decentralization is often discussed in connection with the impact of computerization on the design of an organization. In addition, it is usually dealt with as a special data policy goal, which encompasses the centralization of system development and computer operations. The development of both minicomputers and microcomputers has laid the groundwork for decentralization at the same time that the technology makes an effective centralized coordination possible. In integrated information systems, where telecommunication technology is symbiotically tied to data processing technology, the opportunities for both job autonomy and centralization increase. It then becomes critical to determine in which areas job autonomy is ideal and in which centralization is ideal.

In visions of the future computer society, the bureaucratic organization is discussed. It was Toffler's notion (1970) that work of a more or less routine nature will be performed by large self-regulating machine systems in the super-industrialized society, rendering bureaucratic organizations obsolete. Perhaps this can be expressed in the following way: computers reduce the need for people in bureaucracy.

Under certain conditions, the bureaucracy itself can be built into the data processing system. If Toffler's theses about the relationship between computers and bureaucracy are correct, the disappearance of this bureaucracy would mean that a central obstacle to qualitative as well as quantitative human communications in working life will be avoided (but see below on communication).

Social life and professional life are characterized by an increasing changeability. Changeability and flexibility (as is also true for 'rationality') are honoured concepts in working life. How much change are we able to cope with; how should we prepare ourselves for what are seen as necessary changes? For example, what can upbringing and schools contribute? Our children have needs for completely different social patterns from ours—the generation gap is increasing, resulting in an intense uncertainty about tomorrow.

Combinations of telecommunication technology and computer technology have changed the basic conditions for the local structure of an organization. Which goals do we really want for the localization of work?

- Do we want our jobs situated as they are now, often with long commuter trips from suburbs to the work centres?
- Do we want 'neighbourhood centres' which are located close to residential areas and in which several companies lease premises and open branch offices?
- Do we want to work at home again as in the early days of industrialization? Do we want 'cottage industries'?
- Which short-term and long-term effects can we foresee?
- In which directions do developments seem to be going? Who and what will indicate and select the direction in which we should go? Are we the ones to steer and control; if so, in what direction; if not, how can we play a part in the decision-making?
- Are there appropriate combinations of these options?

Will tomorrow's workplaces be like today's, with dormitory towns and long hours of commuting?

Photo: Jan Håkan Dahlström/Bildhuset

Existing organizational theories suddenly seem a little antiquated. There are neither theories nor empirical results upon which to rely for such matters.

3. Work content

The rationalization philosophy which has hitherto dominated both blue- and white-collar work environments has resulted in a working life which can be characterized by rigid specialization, both vertically and laterally. Work tasks which were created on the lower levels have often been routine and predetermined in detail. But the division of labour principle applies to more than one level. A strict division between planning and operations produces an exaggerated bias in work tasks in both directions.

A new rationalization aid appeared about ten years ago: data processing, which was 'perfect' for the process of removing through rationalization those jobs and tasks which had been created in the Taylorian spirit. For blue-collar workers, computerization has meant freedom from a series of work environments which are dangerous to health, such as those involving lifting heavy weights, noise, air pollution. At the same time, other stress factors were identified. In the newly created jobs, such as the processing operations, psychosocial demands dominate, e.g. analysis skills, attentiveness, swiftness in decision-making.

What was the white-collar situation? First of all, computerization has removed routine jobs from both blue-collar and white-collar workers. Those who perform such work tasks today will in the future increasingly work with other tasks involving somewhat different demands within the same company, branch or even in entirely different sectors. In connection with the first step of the automatization process among salaried employees (white-collar workers) a series of new and fragmented work tasks and occupations that are predetermined (in detail) have emerged, through the application of the same underlying principle of rationalization. However, in the next step of automation, they seem to be disappearing. The long-term trend is less predictable. Present developments towards technically advanced and integrated communication and information systems will most likely imply a fundamental change in the entire office world. More qualified jobs are also being affected, such as those in technological and economic areas (cost estimates and design work, budgeting, production engineering and planning).

Several different methods have been developed to study the changes in working life brought about by computerization. There are broader analyses such as the RAM Project which examine the effects on different areas in the psychosocial work environment (work organization, work content, communication, working hours, aspects of work democracy, health aspects and impact on so-called leisure time). Other analyses have concentrated on and thoroughly analyzed one or more effects on people and their work. Apart from strategy, it has often been possible to attribute observed changes to needs which can be related to the theory of alienation, or further developments of such a theory. The discussion of work content, however, has begun to consider new dimensions, such as: vulnerability aspects in the organization (the possibility of making and correcting errors, interruptions in operation, waiting times); personal integrity on the job; abstraction of task and knowledge; formalization; and standardization.

When more qualified work tasks of an administrative and intellectual character were first rationalized, the impact of the new technological aids on the thought processes and the opportunities for developing creativity, imagination and emotional commitment began to be taken into consideration. In addition, the language of individuals and linguistic development were discussed. Attention has thereby been directed towards processes and conditions which are related to all three concepts of motivation, cognition and emotion.

The concepts of under- and overstimulation seem to be relevant reference points of a socio-psychological nature. Work roles which are characterized either by understimulation or overstimulation arise with the onset of computerization.

At the Postal Giro Service, the RAM Project showed that computerization (a batch data processing system) resulted in work tasks/jobs that can be classified in one of four categories: 'too much', 'too difficult', 'too little', or 'too easy'. Stress arose from all of these, and the organization was therefore subdivided into smaller 'units', with attendant effects on personal relationships. From a crude organizational perspective, the operations departments (bookkeeping) had a number of work environment problems related to understimulation (qualitative and quantitative), while the data processing departments showed signs of problems related to overstimulation (qualitative and quantitative). The two parts of the organization were said to be separated by an 'iron curtain' (compare the organizational effects above).

At the second company studied by the RAM Project, different stress reactions could be identified in connection with the use of an online terminal system. Employees who work at the display terminal for long periods each day feel that they receive too little information, that promotion prospects and personal development and growth in their jobs are too few and that there are not enough opportunities for education and further training. They also feel that they are too confined to their work station, that they have too little contact with other people while at work, and that the physical demands (on the back, muscles, eyes) are too great. Compared with those using the terminals briefly every day, intensive users of terminals more often feel that it is much too difficult to find replacements (such as for when they are on sick leave) and there is too little cooperation at work.

From a labour market perspective, we can imagine a future division of labour resulting in one group of highly qualified, highly educated persons and another group with very low qualifications and a high degree of interchangeability. The gulf between these groups will tend to widen. A third group, consisting of people not gainfully employed or receiving only unearned income ('marginalized' groups or whatever the prevailing terminology may be), will increase, but the size of the increase will depend upon how we conduct the long-term restructuring of our society into a computer society.

Why should work tasks and work environments that can be clustered round one of the two poles of over- and understimulation be avoided in our society? Some important reasons are as follows:

- the risk of sickness and stress (individual level),
- the risk that the labour force will be fragmented (group/organization levels), and
- the risk of marginalization and exclusion from the main stream of society (individual and group).

It is likely that people who enjoy being overstimulated, for whatever reason, will always set their own pace. If the organization in which they work does not allow for this, other channels are open to them. Such individuals may even be better utilized in other positions in our society or in other roles.

From a psychology-of-work viewpoint, it is desirable that a computer system contributes to the creation of work tasks and a work situation that implies as few employees as possible being subjected to the woes of either over- or understimulation. It is the individuals themselves who make this assessment. In the public debate of such matters, discussion has centred on the notion of a 'good enough society', a society of moderation. Perhaps we should begin to talk about 'good enough environments', at work and at home—that is, environments that have neither too much nor too little of various components and qualities (of the respective sectors).

Whether computerization splits the workforce into highly- and poorly-qualified groups, or whether an overall decrease or increase in the level of qualification takes place, will (at least in countries like Sweden) depend to a certain degree on the role given to education and training in that society. Should the educational system be used to force individuals to adjust and conform, or should it be used offensively (for example, to increase the individual's chance of winning influence)? The following points indicate some possible future directions for education:

- Renewed priority should be assigned to systems using so-called recurrent training (cf. the discussions of the 1960s) as a basic pedagogic principle.
- Special training programmes will be needed for certain groups.
- Education should be provided for company employees aimed at developing and improving basic skills required in one's own occupation, and at conveying information about the company's operations and developing expertise about adjoining occupations inside and outside the organization. The aim should be to relate education to the essential objectives of the work involved.
- The content of data processing (DP) training should be made much broader, with more emphasis on societal aspects of computerization. Education should consider both the social and psychological contexts. This applies to the DP training given in compulsory school, and in the occupationally oriented upper secondary schools.
- The training of, for example, psychologists working in corporate health care, or social workers and groups involved in psychosocial work, should offer increased and extensive information on the impact of data processing technology and the opportunities and risks involved in computerization.
- DP education, both in higher education and in compulsory school, should have a technical content and at the same time provide a social science perspective on computerization.

At an early stage in the debate on computerization the pros and cons of people versus computers in their capacity to process information were pointed out. The advantages of computers were described as follows:

- A single mainframe computer can perform the work of several people.
- A computer does not become fatigued and can keep going twenty-four hours a day.
- A computer does not complain about boring, monotonous work. It can be built in such a way that it is immune to a work environment that is dangerous to the health of human beings.
- A computer can perform hundreds of different tasks at the same time, such as conducting individual conversations with a hundred different people simultaneously.
- A computer can quickly store and retrieve information from a

large database, provided that the database has a simple structure and the questions are not complicated.

In their turn, people have the edge in the following ways:

- People understand that special situations demand special measures, which are not always in accordance with the normal routines.
- People often discover errors more easily than computers, since they are capable of understanding what is unreasonable or improbable.
- People are able to take social and ethical factors into account.
- People understand things that are slightly beyond the realm of the work task itself, things that make it possible to handle the work task in a better way.
- People can relate to other people in an intelligible manner.
- People have creative and innovative ideas and are able to grow and to develop skills so that they are able to perform a task successively better.
- People are able to take responsibility for their decisions.
- People are able to key into what is essential.

People often discover errors more easily than computers, by realizing that information is either unreasonable or unlikely.

Photo: Denny Lorentzen/MIRA

The above list of characteristics is closely linked to human intellectual functions. Mankind's entire social contact function, with its emotional components, can be added here. How should work be divided between machines and people? A problem arises when a work task contains some sub-components that are more suited to people and some that are more suited to computers. In such a case, computers and people must communicate with each other. A closer analysis of these problems calls for a knowledge of psychology, and even more so to help provide a firm basis of knowledge for the boundaries to be established when using technology—not least for ethical, moral and psychological reasons.

4. Leadership and the role of leaders

In general, research has not given much attention to the managerial role and to managerial working conditions in connection with computerization. Perhaps this is because so few individuals are concerned, and those who are have had considerable opportunity to influence the process of computerization within their organizations. While it is correct to give top priority to the working conditions of people with unsecured jobs, who are paid the least, have work which is hard, dirty and fragmented, both among white- and blue-collar workers, it is nevertheless easy to fall into a trap here. If a certain management style does not work and the managers themselves are not 'comfortable' with their jobs, the effects will spread like rings on water. A 'bad' manager creates a bad environment for many, many employees.

People with management functions occupy a special position in the development of data processing technology. This technology intrudes and alters their working situation as it does for other occupational groups. Their work environments are directly affected, but such leaders in their turn have a strong influence on the technology through the key positions they have obtained on the basis of their technical and administrative proficiency.

A series of changes both in working life and in society at large have imposed modified demands on management and co-workership. These include:

- the development of democratization in society at large (in family, school, working life);
- an overall raising of the educational level;
- altered organizational ideologies;
- reforms in Swedish working life during the 1970s, such as the

Co-Determination Act, the Work Environment Act and various job security laws;

- the movement towards equality between the sexes;
- technical changes, such as developments in computer technology.

A central factor causing change in the design of the leadership system is developments in computer technology. Systems for planning, controlling and following up operations directly contribute to changes in the same way that the data processing systems for administration and production directly intervene in the work environment both for co-workers and for leaders. So far, leadership and the role of leader have been modified in different ways due to computerization; in our research the organization has shown a tendency to develop into one with fewer levels. Some administrative functions have been built into the system and 'the boss has been made invisible'. Systems which include planning, decision-making and management, tasks often associated with the work of the leader, can just as easily be taken over by leadership groups, working groups within the company or by individual occupants of the posts.

With regard to the development of data processing, it is now time to analyze critically reward and promotion systems. Presently, they are based on antiquated organizational ideologies and traditional role divisions in the home. At the insurance company studied by the RAM Project, the vertical work division was modified and the leadership and the role of leader were deliberately changed with computerization. Difficulties were experienced in the recruitment of management personnel for some sales functions. Higher salaries tended to be less of an incentive. Individuals seen as possessing leadership qualities preferred to forego leadership positions because of the conflicting demands of such jobs (buffer role). Many felt that it was better to be a specialist with a somewhat lower salary. Alternative forms of leadership such as time-limited leadership were discussed as a complement to the present leadership system. In the company's programme for the future there was a demand that a system of rotating leaders be implemented.

In connection with computerization, it is important to take note not only of the horizontal differentiation of work tasks, but also of the vertical equivalent, when designing or modifying the work tasks, and while organizing the work—always keeping in mind human needs and conditions. The reorganization of work can be achieved in different ways:

- introduce other work tasks into the chain of production which are on the same level,
- introduce planning and control tasks that are on another level,

analyze whether some work tasks should be performed by machines,

- analyze which work tasks should be performed by machines and which by people (see above).

Theories on over- and understimulation, empirical research about alienation processes on the lower levels in the organization, critical analyses of the design of leadership and the role of manager, all lend support to the notion that these four strategies should be given more attention when work tasks and job positions are being organized. If applied in a thoughtful manner, computerization may help solve many problems linked to leadership and the role of leader, and thereby with the system and role of co-worker. Current problems often concern the combination of different functions of leadership: social, administrative and specialist roles.

A central function linked to leadership as well as co-workership is the processing of information and dealing with communication in different ways. This will be increasingly true in the future—at the same time that the technical prerequisites are being substantially altered. Many leaders today have a difficult time, partly due to time limitations, in finding a balance between the different functions that are a part of their jobs. Indirectly, this causes irritation among fellow workers. At times, the administrative part suffers, and sometimes the specialist function. Work with contacts, formal standing duties concerning budgets, personnel, etc., takes a lot of time away from the function of specialist. Conversely, it is sometimes true that the leader will slip into—or back into—the specialist role, fulfilling his or her professional commitments and totally forgetting or neglecting the most crucial administrative and social tasks. That a person lives and has experiences from other roles in life—in the family, professionally or socially—should be seen in the future as an asset in working life, not least for persons in leadership positions. It is clear that these roles carry different weight during the different periods of our lives. Many important decisions that affect our professional, family and social lives should be based primarily on solid professional experience and expertise, but also upon experience from the other roles that together make up our lives.

Are not these problems now solved by the computer, which can take over some of the specialist and administrative functions and in this way free a leader's time for the personnel oriented tasks involved in the management of a firm or company, as well as more time for a private life/family life? This is a question of opportunities and risks. The opportunities are perhaps more obvious. Of course, it is possible to change leadership and the role of leadership in connection with computerization (see above). The risks are more complex and they force us to think about current discussions of artificial intelligence—the scientific discipline which examines the relationship between advanced information processing in computers and the human thought process.

5. Communication and computerization

In the RAM Project, display terminal users with public contact, e.g. with customers, complained about contradictory demands connected with a buffer position and felt that the terminal system had intensified the buffer role. The buffer role was most likely reinforced by the faster processing of material which was brought about by the new processing system. The number of customer contacts per time unit increased, which made it more difficult to utilize the higher speed in the production of material for an improvement in the quality of personal contacts. As the work pace at the terminal accelerates, it sets the pace for other functions. Outside demands tend to rise and these are gradually integrated and internalized by the individual.

Work communication can have different connotations. Communication is viewed here from a socio-psychological perspective, mainly linked to working life. The emphasis is placed on the interaction or interplay between people, both quantitatively and qualitatively (see below).

Communication, like language, is a basic factor of group structure. The qualitative aspects of communication and personal relationships take into account such items as the opportunities for problem-solving, creativity, confidence, security, interest, listening, emotional involvement. Even before we became concerned about the impact of computerization on 'man-machine' communication, too little attention was directed towards the qualitative aspects of communication 'man to man'.

The processes of change in connection with computer technology can affect communication directly, analogously with the purpose of rationalization measures or the purpose of the data system. For example, certain types of technology, such as the use of display terminals, lead to direct communication with a computer, at the same time that the underlying system itself is designed by people. Computer technology also had indirect effects. This technology influences many spheres of the work environment (design of organization and work, power structure at the place of work, working hours, etc.). Some of these in turn influence communication between people.

A good emotional climate at a place of work is often the prerequisite for creativity, open problem-solving, etc. If such a climate is present, the content of work, division of labour, working hours, etc. often become less significant. At the same time phenomena linked to these factors initially contributed to the quality and structure of personal relationships.

6. *Some aspects of equality*

Extrinsic equality is closely connected to the question of work division; a central aim is to obtain an equal distribution between men and women of work positions on the labour market and in private life and, not least, of such positions as imply power and influence. In this context, it is of great importance to watch the distribution between the sexes with regard to occupation, professional structure, work division vertically and horizontally.

Intrinsic equality refers to men and women as a whole and as carriers of special cultures. The aim here is that the experiences, values, life styles, approaches, etc. which traditionally belong to women should have equal influence on social life in various aspects; so-called female culture should be utilized and preserved in the formation of our future society. It is important to consider psychosocial aspects in work environment, and the balance between working life and private life and family life and social life.

Data processing techniques affect the working situation of women, partly because the computers take over traditional female tasks and partly because new positions in the field tend to attract males. Furthermore, computers not only affect employment and occupational structure but changes in the work environment as well, e.g., the quality of working tasks and the relationship between work and private life/leisure time.

7. *Considering the future*

The working models (see figures in Chapter 3) used within the RAM Project can also be used as 'thought models' for the structuring of ideal conditions in a computer society. Some important questions are:

- What life conditions (during leisure time, in the family or social life outside what we today call work) do we want in the future?
- What are the ideal characteristics of a good work environment from a broad socio-psychological viewpoint (objective and subjective work environment)?
- What technology (hardware and software) and combinations of technologies can be used to meet the demands and attain the goals stated above?

- What properties in the structure of society, the social frameworks and norms and value systems, promote and create a positive climate for the future development of such a technology?
- In what areas of work and private life should computer technology *not* be introduced?

Knowledge of the behavioural sciences (psychology, education, sociology) combined with other areas of expertise and the experiences and insights of employees at places of work and of persons working at home are all needed to help find the answers to these questions.

2.
Computer technology and its development

1. Phases in the development of computer technology

The golden age of the mainframe computer began in the 1950s and peaked towards the middle of the 1960s. Computer power was to be centralized. Bigger and faster computers with relatively large memory capacities were designed. These computers were used more for purely administrative applications than for scientific and technical applications, which were still quite uncommon. Data processing departments began to appear in large numbers.

This trend turned towards the end of the 1960s, with the appearance of the minicomputer. Although their memory capacity was somewhat limited in the early stages, minicomputers were fully capable of handling a wide range of technical applications and tasks. They were also inexpensive and easier to install and run than the large mainframe computers. Minicomputers became increasingly common in offices and in industry. The decentralization of data processing power had begun. The performance of minicomputers began to be on a par with the now old-fashioned mainframe computers, and their prices were much lower.

The third phase in this process of development was the introduction of the microcomputer. The microcomputer was the main response of component technology to the need that arose in the 1970s for small, decentralized computers which could be built into instruments, machines and other equipment.

The microprocessor at the heart of the microcomputer is:

- highly complex in design, but easily adaptable to a wide variety of tasks;
- mass produced but, with the appropriate software, adaptable to a range of final applications;
- very inexpensive to purchase, but capable of imparting a higher value to the products in which it is used, and of giving them a broader area of application.

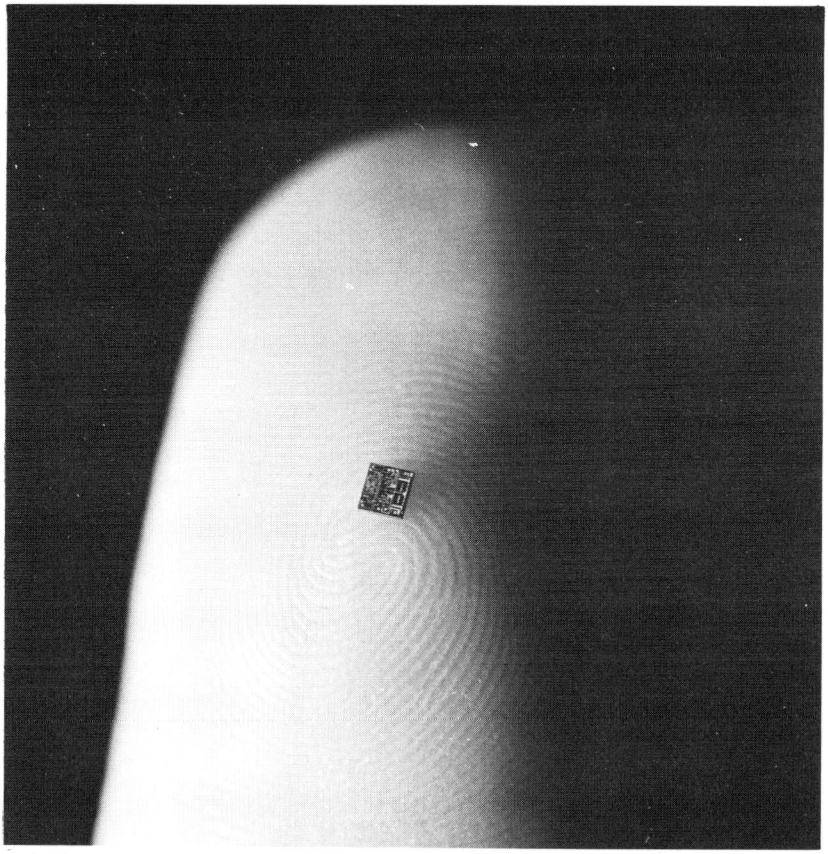

The microprocessor can integrate an increasing number of electronic operations in a smaller and smaller space.

Photo: Göran Assbring/MIRA

Because of the possibilities it opens up for integrating more and more electronic functions while occupying less and less space, the microprocessor is a revolutionary invention. It has proved possible to lower dramatically the price per function in integrated circuits by economical production methods and large batch production. Combined with the semiconductor memory and other electronic components, the microprocessor comprises a complete computer, a microcomputer.

Areas where microprocessors have already been put into service include teletechnology, word processing and graphic displays, data processing terminals, equipment for materials and stores handling, instrumentation and analysis equipment. Today, microcomputers are built into numerous other products such as data collection systems, office machines, cash registers and desktop calculators. They are also used in industrial control, operating and monitoring equipment, making this equipment far more 'intelligent'.

2. Computer technology and the future

What kind of hardware and software will be introduced in the next few decades? The following is an outline of possible future trends.

- It is estimated that hardware will drop by 70 percent in the next decade.* The price of memory equipment will also drop.
- Both storage capacity and processing speeds will increase.
- Equipment will require less service.
- The use of optical reading technology will become more widespread. The human voice will probably become a more common means of data entry.
- Text TV and datavision will have more widespread use.
- The number of computers, mainframe, mini and microcomputers, will increase. Even the smallest businesses will be able to computerize their administration procedures and the operation of their production machines.
- Local networks as well as complex international networks will increase.
- Computer-aided systems will be used more widely for planning, executing, controlling and following up operations, and in compiling source material for decision-making.
- Many end user computer functions will be decentralized.
- Programming and command languages will be simplified, benefiting programmers and making computers more accessible to ordinary users on production lines.
- Computers will improve security and reliability, for example, by means of built-in checks to prevent errors and misuse.
- There will be more opportunities to tailor not only the computer's application and the choice of display and report layouts when installing computer systems, but also whole sets of applications.

Investments in data processing systems used to be more profitable for larger companies. This fact should stimulate the formation of large organizational units, i.e. regional corporate concentrations. But other trends are also noteworthy. The development of minicomputers and microcomputers and the possibilities of long-distance computer-to-computer communication also provide the technical prerequisites for

---

* These are details from the early stages of RAM research project and refer to prices, market conditions, the state of the art etc. from the mid-1970s onwards. They have been left unchanged to preserve the chronology of the research project, and to allow the reader to make a comparison with current market conditions.

The office of the future. *Photo: Gunnar Wählén*

decentralizing company structure. But this decentralization is not an automatic process.

Microcomputers will, however, primarily revolutionize office work. Today, the amount of money invested per employee in hardware used in office production (salaried personnel) is only one-tenth of the amount invested per employee in the manufacture of goods. Demand for data access and data and text processing is growing, as is the need for the preparation of quantitative source material on which decisions can be made. Demands for a more effective use of working time are being heard more frequently, for example, using the telephone conference to replace business trips. All this means that word processing machines, telecommunication equipment and data transmission equipment will be symbiotically combined to bring about a fundamental change in the office work environment. Today it is possible to enter letters into a computer system and 'mail' them through the network. We are on the threshold of the age of the 'paperless office', in which data will be read more and more frequently from display screens.

Transferring data that has been entered into one make of computer equipment to equipment of another make, the compatibility problem, has been—and continues to be—extremely difficult. There are also major compatibility problems between different machines of the same make. This problem exists for word and text processing equipment, for different makes of computer display units and in the area of PCs and communications. It is difficult for all these different systems to 'cooperate' with each other, and there is a risk that tailor-made equipment will have to be developed for each separate routine and system. This trend has been referred to as a move towards the 'gadget office'. Considerable efforts are being made to reduce the scope of this problem. However, in the long term, changes in the field of office information will lead to better integration between the different types of computer, which in turn will be linked to a number of more or less sophisticated personal computers, terminals and what is currently known as word and text processing equipment. Among other functions, this equipment will be able to perform word and text processing.

The progress to be made in the expanding area of communications (see Chapter 3) will enable this equipment to access large quantities of data over networks. The number of communications units will increase and more and more categories of personnel will be using this technology. In time, this trend will bring about far-reaching changes in office work. But an office entirely without paper will probably remain a myth. There will still be good reasons for printing out hard copy in the future.

3. Telematics

Integration between computer technology and telecommunications equipment continues to improve. The different kinds of technical equipment—often occurring in combinations—used as business aids have an effect on the design of occupational roles. The following is a review of equipment that is largely based on telecommunications technology.

There are several systems and concepts in use in this field, and they have been developed from various sources. Functions include the calendar, electronic mail and telemail. The further development of telefax and teletext systems is also under discussion.

- Messages can be sent both within and between companies. Communications equipment can, for example, permit the user to register who has called or paged a person. On returning to his/her work-station, the user can then quickly display this information on the VDT.

- The calendar makes bookings, notes meetings, absences, attendance, etc. This function stores a personal diary that requires constant updating, and allows large meetings to be arranged swiftly.
- Telemail is a process for rapidly sending documents/letters both internally and externally. The document reaches the recipient's in-tray and can subsequently be read and processed on a VDT. Messages/correspondence can be sent, even though the recipient office and its equipment is not online. The in-tray stores all incoming messages. The sender can see on his screen whether the message has reached its destination, and also whether it has been acknowledged.
- Verbal messages: technology allows telephonists to enter and store spoken messages which subscribers can subsequently have played back. Verbal message systems involve sending a spoken message via a system that stores the message until the recipient has time to listen to it (Adriansson *et al.*, RDF, 1982.
- Telefax and teletext are kinds of 'Supertelex' (thirty to fifty times faster than traditional telex machines). Telefax is used to transmit diagrams and pictures, and teletext for transmitting text. Teletext appears to be expanding to include many different types of communication, for example, data and graphics. Ideas have been put forward for combining graphics and text displays with the human voice in one and the same document (*Datavärlden*, 30 September 1982).
- Teledata is defined in a Swedish Government Report (1979:69) as transmission via the telephone network of information entered in one or more computers to a terminal that may be a modified TV set. Five functions have been defined in the main use area of teledata technology. These include message/electronic mail functions. The other four functions are as follows (Junberger, 1983):
  - First there is an information system that contains up-to-date information from various sources. The teledata base is also a database that is particularly suitable for current information and fresh items, continuously updated for mass consumption. In its basic form, the system can perform certain transactions. However, these cannot be online processed, i.e. in the form of time-critical operations. Instead, these are based on transmitting either a freely formulated text, or a 'reply card' on which a physical recipient can act. Ordering brochures is a good example of this.
  - The second function is transaction oriented. An ordinary teledata terminal is the interface with process computers with

dedicated functions such as bank services. Gateways or interfaces for external computers are the bases for these functions. In all cases, available communications networks are used—telephone, data and satellite networks. These external data applications are already in use.

- The third function is education oriented. The simple interface—the television screen, colour and graphics options combined with an alphanumeric keyboard—offers the same opportunities in school classrooms as do considerably more expensive computer systems. Teledata technology takes education to the pupil instead of the reverse. The Open University in England has a great deal of positive experience of this way of disseminating knowledge.
- The fourth function is leisure and entertainment. Today the amusement arcade is computerized. Space invaders and Star Wars are common games and have taught technology to thousands of young people. These programmes, which are termed unhealthy by sociologists, may be superseded or supplemented by more intellectually oriented products such as chess and other quality games. Here, the parallels with programmes modified for educational purposes are obvious.

3.
A research project on the psychosocial aspects of computerization

1. Background and purpose

From the mid-1970s to the mid-1980s I conducted a major research project at the Department of Sociology, University of Stockholm on the impact of computerization on the work environment. The project was called 'The Effects of Rationalization on the Work Environment and Working Conditions in Some Salaried Employee Groups' (the RAM Project). [RAM is an abbreviation of the key words in Swedish, *r*ationalization R, work environment (*a*rbets*m*iljö) AM.]

The main purpose of the project was to make an analysis of the psychosocial problems in the work environment and examine their relationship to changes which occur through the rationalization of administrative procedures, mainly those involving computer technology.

The first RAM Project, referred to here as RAM:1, consisted of two sub-projects:

> Project I: a study of a government-owned company, the Postal Giro Centre (see Chapter 4);
>
> Project II: a study of an insurance company, Trygg-Hansa, Stockholm (see Chapter 5).

Thus, from the technical viewpoint, two main types of system were studied: batch-type systems (batch data processing) and online systems, using display terminals. Each system represented different conditions for the structuring of work environments, which may thus be said to imply different operating assumptions for work environments. In the final phase of the RAM:1 Project a new, larger research programme, *Computer Technology and Working Life* (Bradley, 1979) was produced. This project is referred to here as RAM:2.

Proposals for continued research involved two strategic choices. Firstly, research was intended to focus on the study and analysis of microcomputer systems— the third phase in technological development. Secondly, the research was to concentrate on three problem areas, or three areas which, on the basis of previous research, were considered to be key problems in terms of the development of computer technology. This research is still in progress, but the experience gained from one of the sub-projects is summarized in Chapter 6.

The RAM:1 Project had its origins in the inventory of problems in the psychosocial work environment commissioned by the Swedish Confederation of Salaried Employees (TCO) several years ago, and which is reported in '*Work Environment and Salaried Employees*' (Bradley, Börjeson and Lundgren, 1974).

2. Problems

The RAM:1 Project attempted to find answers to a number of questions, including the following. The problem areas listed here relate to the concepts that appear in Figure 2.

- How does the introduction of computer technology affect the general organizational structure of a company?
- Are radical changes taking place in the content of work tasks?
- Does work content change in terms of the meaningfulness of the job, its utilization of human resources, and the degree of independence/self-management and responsibility (job autonomy) it offers the employee?
- What is the effect on the real influence people have on their work and on their interest in exercising such influence?
- Are new categories of occupation created and, if so, what are their characteristics in terms of the individual's ability to influence work methods, work pace and contact with other people? Do these new occupational categories create a freer or more controlled work situation?
- Does the organization's internal communication pattern change? Is there any effect on the amount of contact and cooperation between people at work—in their own department, between employees with similar work, and between different departments in the same organization?
- Is the 'buffer role' of many employees weakened or strengthened (role ambiguity)?

- What changes are there in the management system, the patterns of promotion and attitudes towards personal development at work?
- What changes are there in the co-worker system?
- Are new social groups emerging in the labour market? Will the introduction of computer technology eradicate the traditional distinction between white-collar and blue-collar workers? Will new groups of white-collar workers emerge within the white-collar sphere, for instance?
- Will there be any change in the differences between the status of men and women in terms of working conditions?
- Are changes taking place in working hours, e.g., shift work and overtime?
- Are salary structures affected?

Both objective and subjective (experiential) changes related to these problems were studied. Also an analysis was undertaken of changes in life conditions outside the work sphere itself, both objective changes and changes in experiences and attitudes related to people's private lives and leisure time.

3. Methods

Table 3.1 provides an overview of the data acquisition and information analysis methods used in the project. The terms 'objective' and 'subjective' work environment used in this table recur in the models used for the project.

In the early stages of the research project qualitative methods were used, for example, interviews, the analysis of documentation and a study of relevant literature. We then transferred to more formalized methods for data gathering and analysis, i.e., both descriptive and analytical statistics. In the later stages of the research process, however, we returned to the qualitative method for study, interpretation and feedback. The choice of method was dictated by the relevant phase in the research process, in other words by the type of problem we wanted to highlight.

It was considered very important to start the work with an exploratory phase. In 1974, when this project was started, few socio-psychological studies of the impact of computer technology on the work environment had been completed. The research results that were available dealt mainly with conditions in other countries, where circumstances in the labour market were different. Research often

Table 3.1 Summary of data acquisition and processing methods used in RAM project

	Objective work environment	Subjective work environment
Data acquisition methods	Behavioural observation Internal documentation, e.g., company statistics, minutes (from works councils, board meetings) Interviews with management representatives, trade union representatives, AR men/women[1]	Interviews, semi-structured Questionnaires, mainly multiple-choice answers Data on short-term absenteeism, staff turnover trends, etc. The diary method
Data processing methods	Qualitative analysis of administrative material Analysis of available documentation Descriptive statistics	Qualitative analysis of interview material Descriptive statistics Multivariate analysis techniques

Note 1: People involved in the company's work for administrative rationalization (AR), work groups or the head of the department in the company that dealt with AR work connected to computer-based information systems.

focused on blue-collar workers. In addition, most white-collar worker studies concentrated on the work environments of computer programmers and systems specialists.

Moreover, as anticipated, the use of a strategy based on 'exploration' in an attempt to highlight previously unknown changes in people's working conditions proved to be fruitful. Naturally, formulating hypotheses would have been an alternative approach. There are in fact no major differences between hypothetical-deductive research and critical sociology-psychology. The problem areas of the type we explored in the RAM project are based upon non-directed types of hypotheses. The relationships between the concepts are also hypotheses. The 'hypothesis formulation' is reproduced in more concrete form by the arrows in the models (Figures 1 and 2).

It was possible to translate the theoretical models to their empirical counterparts, both in sub-project 1 (Post Giro) and in sub-project 2 (Trygg-Hansa). Priority has been given to analyses and processing that are important in terms of current problems in working life. Subsequently, various attempts were made to examine in more detail 'sections' of the models and to select certain strategically dependent variables (see Chapters 4, 5 and 8).

4. *Theoretical models and concepts*

There is both a physical and a psychological side to the work environment. The concept may be defined briefly as follows: the work environment includes everything that affects the individual in his work, and everything that is affected by the individual in his work. When approaching work environment problems, it is important to adopt a broad perspective, so that there is no risk of overlooking factors that affect the relationship between society, the company and the individual.

The work environment at different levels

In the RAM Project, a broad conception of work environment was used. This was an important point, not least in terms of finding ways of rectifying shortcomings in the work environment. The work environment can be separated into three levels: the level of the society, the level of the company and the level of the individual. At the *level of the society* the individual is affected by societal structures—political, economic, ideological (religion and culture), technical (e.g., the degree

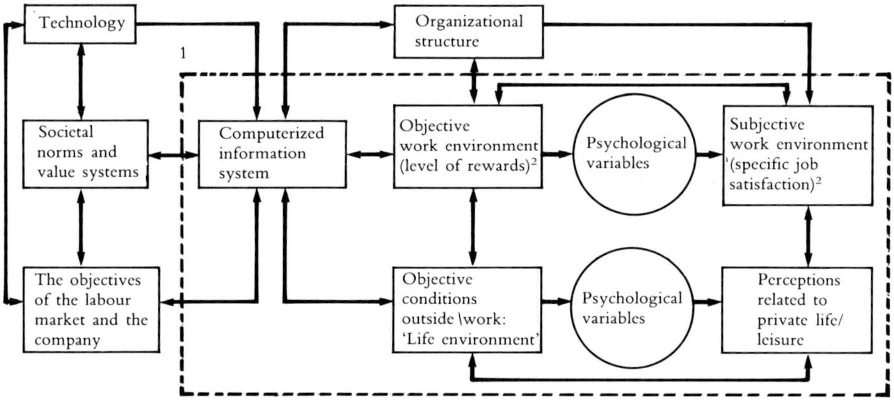

Figure 1. Computer technology and the work environment (Bradley's outline model)

Keys to symbols:
→ Influence
↔ Two-way influence

Notes: 1. The outlined field states the range of variables on which the RAM Project concentrated.
2. More detailed specification of the relations between objective and subjective working environment is given in Figure 2, which contains examples of these concepts.

Figure 1 is reproduced in Bradley (1977) and Bradley (1981) with more detailed explanations of the concepts and relationships.

of industrialization) and social. At the *level of the company/organization*, the individual is affected by a number of interdependent factors, such as ownership structures, organizational structures, the conditions of power and influence, wage levels and pay systems, the content of work, working hours, rationalization and physical and spatial conditions. Each individual has a unique perception of his/her work. Factors such as the individual's inherited characteristics, state of health, experiences, knowledge and needs are at the *level of the individual*.

There is interplay between the factors at these three levels. The individual is not an organism without a will of its own, or an object that is passively affected by different factors in its surroundings. People are products of their social environment, but also have the power to influence and mould their environment themselves. This

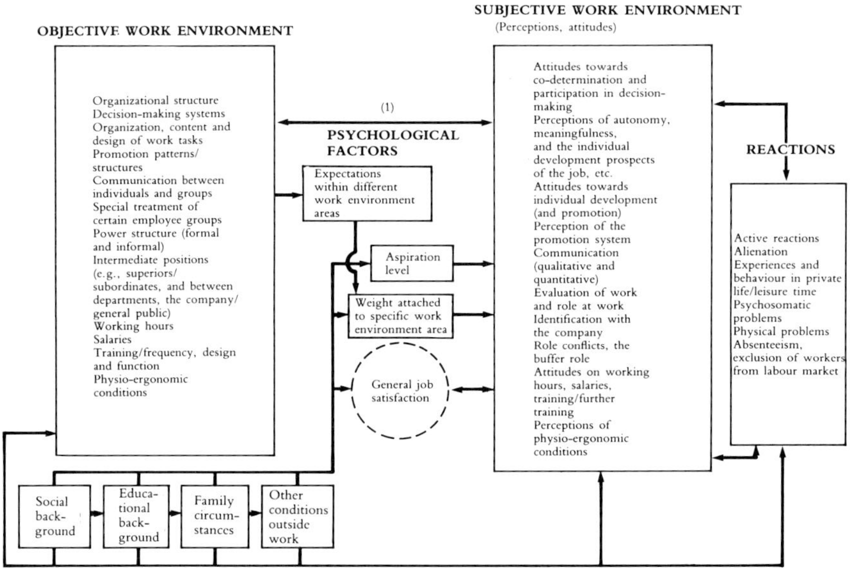

Figure 2. The relationship between objective and subjective work environments (Bradley's model) (Sources: Vroom, 1964; Dahlstrom *et al.*, 1966; Bradley, 1972 and Bradley *et al.*, 1974)

Key to symbols:
→ Influence
↔ Two-way influence

Note:

Figure 2 is reproduced in Bradley (1977) and Bradley (1979) with more detailed explanations of the concepts and the relationships between them. Figure 2 further specifies the relationships in Figure 1.

interaction between individual and social environment is within the purview of social psychology. In recent years attempts have been made to develop the individual's power to bring about change by promoting various ideas such as 'the desire to bring about change', 'training in positive thinking' and 'developing yourself and others'.

According to this argument, work environment problems must be tackled simultaneously at different levels—by political action, for instance legislation, and also by bringing influence to bear on work content, wage forms, promotion systems, organizational structures, working hours, rationalization and other aspects. It is important to aim at introducing preventive measures without neglecting the changes that are important for the individual here and now (see Chapter 9).

As an example of the relationship between these levels, the competitiveness among individuals to be found at a number of workplaces may be said to be a reflection of events at other levels. In the last analysis, it reflects the type of society in which we live. Between these two levels there is a level where the company's organization and its pay and promotion system encourage competitiveness between individuals. The foundations of this competitiveness are laid at school and strengthened by factors such as the system of school credits which in its turn is a product of the structure of society outside the school.

Objective and subjective work environments

As described above, the work environment can be approached from different levels and from different perspectives. It may also be divided into two categories: the objective and the subjective work environment. Factors in the *objective work environment* include the company's organization, the decision-making system, the promotion system, the framework of working hours, pay, intermediate positions and physio-ergonomic factors. A number of these factors are interrelated. The number of levels in the company organization—whether it is a lateral organization or if the company has a hierarchical, pyramidal structure, how the departments are related to one another and what communication channels exist—are aspects reflected in, for example, the promotion system. This may involve the number of job promotions, the qualification requirements for these jobs, their internal interdependency, reporting channels and other aspects of the promotion system. The *subjective work environment* includes employees' attitudes and perceptions and their overall job satisfaction, covering attitudes towards employee participation in decision-making processes, job autonomy, the meaningfulness of the work, the opportunities for development afforded by the work, attitudes towards personal development, role ambiguity/the buffer role, attitudes towards pay and working hours, etc.

A company's structure, whether it is strongly bureaucratic or based on a form of autonomy, is reflected in a number of ways, for example, in employees' attitudes to industrial democracy and having influence in decision-making processes at work. The actual design of work tasks, for example, certain repetitive movements, excessive supervision, constant interruptions of a job that requires attentiveness, constant new personal contacts, is reflected in the way people perceive the content of their work, and helps form different reaction patterns (see also page 35). In its turn, this type of perception has an effect upon the objective work environment.

Figure 2 illustrates the concepts in more detail. A two-way arrow links objective and subjective work environments. The underlying concepts are most easily understood through examples. The term 'psychosocial work environment' is used to signify the course of events or the process that occurs when objective factors in the environment are reflected in the individual's perception (either positive or negative) of work and the conditions of work. Its essence is the interaction between the environment and the individual.

Thus work environment factors exist at different levels—the level of society, the level of the company and the level of the individual—and they interact with one another. But there is also a distinction between the objective and the subjective work environments, which also interact.

Expressions like 'job satisfaction' and 'job fulfilment' are popular terms to summarize the different components that constitute the subjective work environment. These are placed on the right of Figure 2. However, the import and significance of these components may vary from one person to another. To a considerable extent, the experience of work can be explained by what actually happens in the objective environment. But there is a complex interdependency here. Our disparate backgrounds, aspirations and basic values and, in addition, any changes in these factors, all affect the way we perceive our environment.

Naturally, environmental problems can be discussed using other terms, but the phenomena in the work environment represented in Figure 2 can be analyzed in relation to both blue-collar and white-collar workers. What is then the most important factor may vary among different groups of employees, different sectors of the labour market, different trade unions, etc.

Why should the objective and subjective work environments be kept separate? If only terms such as 'satisfaction', 'adaptation', 'perceptions', 'attitudes' are used in the debate on the work environment, there is a risk that we will lose sight of the underlying causes, that we will not get to the root of the matter and uncover the controlling mechanisms behind these experiences.

This is not to say that our feeling for or our enjoyment of our work is unimportant—having positive feelings towards our work is vital for us, and it is important to pose questions about satisfaction and attitudes. The work environment is best evaluated by the person who is exposed to it. But at the same time we must use a variety of means to supplement this picture with descriptions of the objective parts of the work environment, e.g., the design of the salary system, the structure of the organization, the design of technology, etc., that will give us a far better basis on which to discuss changes in the work environment. Figure 2 summarizes the way in which different factors at work are reflected in people's experiences and feelings at work, and indicates some of the effects, and the ways that reactions express themselves.

The effects of a poor work environment: reactions

Reactions to an adverse work environment take various forms. A common reaction is to show signs of what is usually referred to as *stress*. Other reactions are to experience disappointment, a sense of powerlessness and alienation from one's work. But reactions may also be expressions of a positive attitude, a desire to make changes and improvements, and various reactions that indicate well-being.

When people experience prolonged and repeated stress, a number of symptoms persist, symptoms of what are known as *psychosomatic illnesses*. The most common are stomach ulcers, high blood pressure and vascular diseases. These may be called 'long-term effects'. (See Chapter 7 on stress and computerization.)

In recent years, various psychosocial factors have been observed in relation to *industrial injury*. This term now also includes injuries caused by socio-psychological factors in the work environment. When reporting industrial injuries, one is asked to give details of any contributory factors. In addition to the technical design of the job/workplace and safety equipment, the organizational design of the job/workplace is also mentioned as a factor, as are work methods and supervision/monitoring.

By and large, concepts of *alienation* originate in the theories of Marx. Subsequent research has uncovered a number of factors that have a significant impact on what one experiences in one's work. These include technology, work allocation, social organization, e.g., bureaucratization, and the economic structure. In the discussion on which factors cause the feeling of alienation at work, three basic human needs have been put forward, and a description of what happens if these needs are not met:

The need for self-determination and job autonomy	→	Powerlessness		
The need for interesting and meaningful work	→	Meaninglessness	→	Alienation
The need for association and identification with the group	→	Normlessness Isolation		

Alienation can be described as an emotional condition that is caused by a fundamental disappointment in terms of one's need for self-determination—job autonomy, interesting and meaningful work and a place in society. The disappointment of not having these needs met creates in people a feeling of powerlessness, meaninglessness and normlessness. The individual directs his involvement away from the work as such, and seeks to satisfy his needs in his leisure time through various kinds of consumption. One makes a passive adjustment to work. Terms such as 'a low level of work commitment' and 'an instrumental attitude' are used to describe the individual's reactions. The work loses all its intrinsic value and becomes important only in terms of what it can provide for leisure activities, i.e., salary benefits and working hours of the job. But it is also possible to have an active reaction. The individual can attempt to change his work situation in a number of ways, either on his own or by uniting with others, for instance in trade union activity. These actions may be aimed at the different levels mentioned above. When their real intentions are fulfilled, new labour legislation and labour market agreements can create better conditions for employees to take positive action and bring about changes in working life. The extent of the impact of this process of adaptation, which is expressed in passivity and resignation, may then decrease.

Changes in employees' lives outside their work—primarily changes in their families—may also affect their attitudes. Greater alienation at work and instrumental types of behaviour can be caused by putting priority on one's family or private life. The relationship to work then becomes purely an economic one (Goldthorpe *et al.*, 1971). Association with one's family and children may also be a source of inspiration, and may revitalize the conditions that apply in working life.

Absenteeism from work (for which the underlying reasons are very complex) and the phenomenon of marginalization or *exclusion* from the labour market are also expressions of a poor work environment.

Finally, shortcomings in the work environment may express themselves in a way that causes problems in the employee's *private life*. Phenomena at work are directly reflected in people's relationships with their family and their private lives in general.

Summary

The work environment has been considered in terms of the following: the different levels; the objective-subjective work environments; the interplay between working life and private life; and the life cycle perspective.

The work models of the RAM Project are illustrated in Figures 1 and 2. These diagrams show the relationship between the objective and subjective work environments, the different effects and the ways phenomena are expressed.

The content of some of the concepts expressed in the models may be summarized as follows. The '*objective work environment*' refers to a selection of areas of work that are germane to large groups of salaried employees (Bradley *et al.*, 1974). The selection of factors in the work environment is based both on a survey of literature within the project, and on interviews conducted during the planning phase with people working on rationalization issues, and who are involved in or affected by these issues in companies, administrative authorities, institutions and trade unions. The definition of the factors was further refined through a comprehensive series of interviews with employees in the organizations studied in the project (the qualitative phase).

The '*subjective work environment*' consists of perceptions and attitudes related to corresponding sets of factors in the objective work environment. The subjective work environment is closely linked to the concept of job satisfaction.

'*The organizational structure*' includes the methods used to allocate work, the basic material for decision-making and organizational aids. The concept of organizational structure is closely related to the concept of the work environment. The design of the organization is an important factor to consider when studying the relationship between electronic data processing (EDP) and the work environment. It can also be seen as part of the work environment. However, a study of the relationship between EDP and the structure of the organization was not the main concern of the project.

'*Objective conditions outside work*' refers to behaviour and consumption, the conditions that prevail during the hours spent away from work. These may be affected by change related to the use of EDP at work. Certain attitudes, values and experiences related to private life/leisure and family life were also analyzed against the background of the introduction of computer technology into the individual's work.

'*Psychological variables*' is a general term covering a number of intermediate, psychologically relevant variables such as the level of aspiration and the weight attached to specific work environment areas. These variables are essential if one wishes to understand and explain the subjective work environment and its dynamics, and the perception of the conditions that govern our lives in general.

With reference to these figures, the following points are important:

- Objective conditions outside work should be included in discussions on the introduction of computer technology. We should count on both indirect effects of computerization in working life and the computerization of functions that relate directly to our private lives. Goods and services are produced both in working life and in our private lives, but under different conditions and with different economic bases.
- We should be more aware than before of psychological variables and psychological phenomena (attitudes, aspirations, expectations).
- We should pay attention to the two-way arrows in the figures. For example, it is important to recognize 'reactions' (ways of reacting) that can help to change conditions at work and in our private lives that we are dissatisfied with today and that can strengthen the things we feel work well, both as individuals and collectively.

The work models used in the RAM Project can also be used as theoretical models in discussing the structure a computerized society should have—see the two-way arrows in Figures 1 and 2.

5. Concluding remarks

This chapter has described two psychosocial models of the relationship between computer technology and human beings, which form the point of departure for the research. They are broad-based models and can be used for deeper study of aspects of more general models.

The models illustrate a view of working life and its processes. However, there is a risk that the same phenomenon that occurs in working life in general—the fragmentation of knowledge—will also occur in the field of data research and research into the consequences of computerization. Integration is easily lost and it becomes increasingly difficult to gain an overview of the research results in terms of computer technology and various combinations of new technology. Parts of the reality can be separated out and described in great detail—effects can be shown—but the force of these effects in the interplay with other parts and 'fragments' is often uncertain.

Atomization versus the holistic approach has always been one of the dilemmas of research, but becomes a more serious consideration in some disciplines. Lars Kristiansson has the following to say on 'models and reality':

> 'It is not only at restaurants and in the forest that people use models of reality. Our entire western society attempts to capture the past, the present, the future, ants, humility, stones and love in oversimplified models. We use everything from mathematics to highly-emotive brush strokes as model languages.
>
> All models, whether they are described in the language of mathematics, the symbols of cartography, or the words of a poet, have limited validity in the light of reality.
>
> Sometimes more than one model is needed to describe reality. In a foreign harbour, it may be good to have both a street map of the town and a sea chart. They are complements to one another, but when they overlap, they must not be contradictory.
>
> The mathematical model does not compete with the poetical. They complement one another in our attempts to find answers to questions about life, about how it can be improved and, preferably, how it can be given meaning.' (Kristiansson, 1982, pp. 14, 15 and 16)

4.
Experiences from Company 1: The Postal Giro Centre

Various state agencies/public utilities were contacted to ascertain if it would be possible to run the project in them, and whether they were suitable organizations. This involved an examination of various documents, and their rationalization surveys, systems descriptions and organizational descriptions/plans. The organization had to be interesting from the data processing technology viewpoint; have intelligible production procedures and administrative structures; and possess broadly applicable areas of activity which would produce generalizable research findings. A government-owned company, the Postal Giro Service (Postgirot), with approximately 5000 employees, all of whom work in the Stockholm area, was finally chosen for the RAM Project's first field survey.

1. Computerization in the Postal Giro Centre

Ever since it was founded in 1925, the Postal Giro Service has used mechanical equipment. Adding machines, accounting machines, stamping and sorting machines have been essential complements to the work carried out by human beings. In the history of the company, the period from 1947 to 1967 is known as the 'punch-card era'. The first computer was installed in 1962. One of its first tasks was to take care of the growing volume of payment-in punched cards, which had been introduced in 1957. In mid-1963 the accounting of payment-out cards was transferred to new equipment. The first step towards automated bookkeeping was a partial automation, completed by September 1967. At that time, all accounts were transferred to EDP. The EDP system described below is the second step towards automation in this organization.

Background

The decision to start the EDP project was preceded by an extensive survey, which examined possible alternative systems. An important strategic decision was taken when the preprinted form was retained as the data carrier. The organization's assessment was that the form would be used as a data carrier for at least ten more years, partly because society was not yet ready for a rapid transfer to a widespread payments system using neither cash nor forms. A system was therefore chosen that was markedly different from the direct-link terminal system.

Objective

The EDP system was a new, central system for collecting, processing and transmitting data. It was introduced in two stages. The system was expected to result in manpower savings in the order of 900 man days, or, expressed in kronor, 18·0 million in 1975, and 42·0 million in 1977*. New forms designed for optical reading would also simplify the company's customer routines.

Technical description

The EDP system was based on optical reading (OCR) of incoming forms both from accounts customers and branch offices. (Mechanical input was made possible by adding supplementary codes to the information on the document in an optically readable form.) This coding replaced the former hole-punching process, but it was also essential for the mechanical sorting of documents.

The OCR machines transferred the data to tapes used for data input to the central computer. The machines also sorted the documents which had been read. All documents were micro-filmed when the data were collected in the OCR machines. The documents then accompanied the statements of accounts as verification. The film, which was used in correction routines and as documentation for any complaints, took up far less storage space.

The EDP system required the basic investments shown in Table 4.1. There were additional variable costs, directly attributable to the EDP system, of approximately SEK 10 million per year.

The mechanical processing of all data consisted of a main accounting system and a main auditing system. Neither the accounting system nor the auditing system were fully developed in the initial stages.

---

* Approximate prices in the mid-1970s.

Table 4.1 Investment required for EDP system

	SEK millions*
OCR equipment	30.0
Coding machines	8.9
Microfilm equipment	0.8
Conversion of premises	2.3
Systems development	2.3
Investment in forms	1.0
Information	1.0

* Approximate prices in the mid-1970s.

Both systems were extended in the next phase, when the main systems were modified to accept the new data acquisition technology. Some sub-systems were also adapted to the main systems. Consequently, the mechanical changes were negligible. These changes concerned the organization of work routines.

Work routines before and after the introduction of the EDP system

The following brief description of the work routines before and after the intoduction of the system are taken from the systems description and other internal documentation available at the time the investigation was carried out.

The Bookkeeping Departments

The new data system involved widespread changes in these departments' routines. Before the system was introduced, incoming forms were processed manually. The routines involved extensive sorting, examination, cutting and filing of the forms. Finally, all transactions relating to the same account were totalled on an adding machine. The totals of both deposits and withdrawals for each account were punched on paper tape that was fed into the computer. Most of these steps disappeared with the advent of optical data acquisition, which requires little manual processing of incoming documents. The data then go straight to document coding. The transfer of data to tapes and the sorting and microfilming before filing are now carried out with the aid of OCR equipment. The so-called CMC correcting routines were introduced to process documents that had been rejected by the OCR machines. Formats and list files were replaced by microfilm.

The Investigation Department

The data lists were replaced by microfilm. (No other information is available here on changes in the work routines.)

The following sequence of pictures describing the Postal Giro Centre's accounting process at the time of the RAM project's survey of the organization is taken from *A History of the Post Office Bank, the Post Office Savings Bank and the Postal Giro (1884–1925–1974)* by Kurt Samuelsson (1978).

The Auditing Department

The following changes took place: the payment forms were included in the credit check system. After the payment form was redeemed, it was sent from the cash desk to the Auditing Department. The payment forms were totalled on an adding machine tape. The totals of the adding machine tapes were then transferred to punched tape. The introduction of the computer system forms meant that the payment forms could be read by an OCR machine, and the punch card process disappeared. Data acquisition became simpler and correction routines disappeared. The data input described above also disappeared, replaced by direct terminal transactions.

The Accounts Department (personal accounts)

A terminal system with VDTs had recently been introduced. The system's ten VDTs were linked online to the computer and were used for checking balances. The response times were extremely short. The terminals were intended both to provide a better service to the cashiers when they called the Accounts Department, and to reduce the incidence of overdrawn accounts. Before this terminal was introduced, the clerk used to leave the telephone to look up this information in the balance lists.

2. Research design and methods

Only mainframe computers were in use at the Postal Giro. A batch-type system was in use (with some operations performed by the data processing system and others by the employees). The staff had no direct contact with the computer. The work materials often consisted of data listings instead of, for instance, texts displayed on the terminal screens.

Data were gathered from internal documentation (company statistics, minutes from meetings, etc.), interviews with staff who were specialists

The Postal Giro Service's bookkeeping operation in modern times (1975)

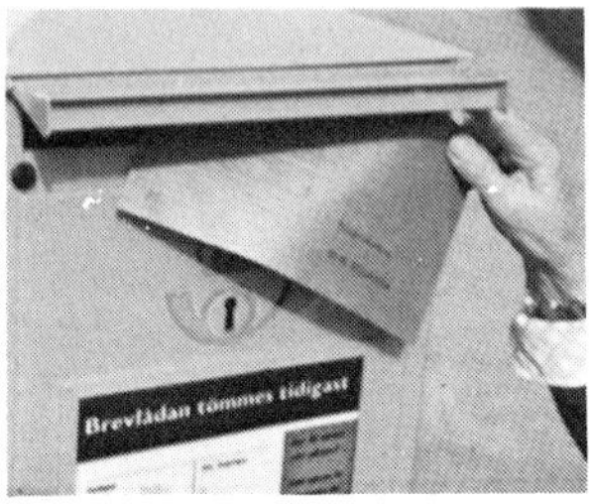

1

1. **The brown envelope**
The symbol of the Postal Giro is put into one of the country's 38,000 postboxes. It is said that the Post Giro is open twenty-four hours a day. Payments in are also made in post offices and postmen serving rural areas collect post giros on their rounds. All this happens on day 1.

2. **The post arrives**
Day 2 – the post room. Work starts here at 06.30. Last night's harvest flows in and is handled automatically on conveyor belts. 60,000–70,000 brown post giro envelopes containing some 500,000 assignments arrive daily and are first sorted by hand and quickly disappear on the conveyor belt to the four bookkeeping departments. Even more work has come from the post offices, and the yellow envelopes containing payment-in forms (the so-called missives) are processed in a different department.

3. **The 'missives'**
The sorting centre handles missives totalling a sum of 500,000 payment-in transactions every day. Here, the forms are checked and the pre-coded forms are separated for processing in optical reading machines so that the computers can start as early as 07.00.

4. **Bookkeeping**
Extensive sorting and checking of the old forms, and coding the new forms for optical reading is carried out during the morning in the bookkeeping department. The current introduction of new forms is an extensive project called G70, and is the visible sign of rationalization aimed at extensive automation of the bookkeeping procedure. But some manual operations cannot be avoided.

2

3

4

5

6

7

8

5. **The coding machines**
A coding machine prints the amount and the type of transaction, etc. on the bottom free field of the form. At the same time, the total is checked. A minicomputer in the coding machine gives the alarm if, for example, the control number is wrong or the wrong amount is keyed in.

6. **Optical reading**
Some material is sent straight to the Postal Giro Centre for optical reading, the so-called REI hall, as 60 per cent of all payment-in cards are pre-coded. REI stands for Recognition Equipment and it is in the eleven OCR machines (Optical Character Reading) that all the information is converted to data language. Here, the material is automatically optically read from 07.30 to 12.00, and is sorted until 16.00.

7. **Speed**
What happens in the OCR machines? These are very high-capacity machines which can read over 2000 forms a minute. Each form is stamped with the date and is also micro-photographed. The process is too fast for the eye to follow. The form moves at a speed of 40 km per hr through the machinery before being sorted into different boxes. During this procedure, the machine has transferred the data to the magnetic tape that is sent to the computer room's IBM machines where it is read into the 'memory'.

8. **The microfilm**
But the material in the REI hall cannot be further processed before a microfilm has been developed and the tape checked and reconciled. These data are read on monitors. The microfilm is the only filing material that the Postal Giro retains, because the forms must be sent on to the customers. The Postal Giro has its own microfilm laboratory fitted with the new G70 system, which can develop some 3 km of film every day. Two copies are made at the same time. There is room for forty forms on 1 decimetre of film.

9

9. **Corrections**
The forms that the OCR machines could not read completely are rejected (if, for example, part of the code strip has been defaced or changed). These rejects must be punch-coded or 'corrected' manually. In a separate computer run, the machine reads the fields in the code strip that can be read and puts question marks next to the other fields. In the correction process, each form is individually checked against the data from the magnetic tape on a display unit, and an operator can add the missing data to the form. When the form is completed, and the data entered on a separate magnetic tape, it is sent on for computer processing and reconciling.

10. **Final check**
If the computer signals that something is still not right, computer printouts of error lists are requested and, using the microfilms, the error is traced back to the original form. This work continues until the day's transactions are checked as correct. The total giro cycle contains a series of checking operations to ensure optimum security for the Postal Giro Service's customers.

11. **The IBM room**
In addition to the OCR machines mentioned earlier, the Postal Giro computer room has three IBM 370 main units, thirty-four disc drive stations and twenty magnetic tape stations, four punch-card readers, three punch tape readers and twelve printers.

12. **Disc stations**
All details (account number, addresses, balance and turnover, etc.) necessary for the bookkeeping are registered on disc. The actual bookkeeping operation uses these data. Each station contains data for the day's bookkeeping for one-quarter of the accounts. Each so-called diskpack contains 100 million characters. There are twenty disc sides in this kind of pack and the transfer rate is 806,000 characters per second. This equipment can retrieve data on a balance, etc. extremely quickly.

10

11

12

13

16

14

15

13. **The printing department**
The twelve fast printers in the Postal Giro computer room have a considerable capacity. In the space of a couple of hours every day, an average of 125,000 account transactions are printed out, i.e., the reports that are sent to the customers on day 3. The speed of these machines is impressive; they write 2000 lines a minute and one line can contain up to 150 characters.

14. **Tape storage**
The Postal Giro Service has a large tape storage area with both its own and other organizations' magnetic tapes. By exchanging computer tapes either directly with the customer companies or through a data service company, the Postal Giro can provide an extensive service to companies, organizations, and other major customers.

15. **Sorting machines for withdrawal notices**
Via the bookkeeping departments, where the forms and verifications are checked and collated manually with the account withdrawals, the circle is quickly closed during the afternoon when everything is put into envelopes. Now other machines take care of the greater part of the sorting operation. A seven-character code is attached to the addresses. Each of these machines sort approximately 25,000 envelopes per hour. The time is now between 18.00 and 20.00. The working day is not finished before all material has been processed.

16. **The circle is complete**
The customer receives his withdrawal notification on day 3 and can then see the complete accounting. If the customer has any questions about the accounting, modern technology will now rapidly retrieve any data requested from the Postal Giro investigation or auditing departments.

in the issues being researched, 'expert interviews', semi-structured interviews with employees from the company and questionnaires consisting of fixed alternative answers. A cross-sectional method was used in both the loosely structured interview phase (qualitative) and the more structured phase of the questionnaires (quantitative). All the subjects of the survey had had some involvement for at least one year with the computer system being studied. The different phases of the investigation were carried out in 1975 and 1976.

Three instruments were designed: a semi-structured guide for interviews with experts (see above); a semi-structured guide for interviews with employees; and a final, detailed questionnaire, 'Work Environment and Rationalization' (approximately 400 items). The instruments show three possible steps towards an evaluation of the EDP system. They are examples of methods for evaluating the effects on the work environment in the broad sense, relating both to planned EDP systems and to systems that were already in use.

The instrument used in the quantitative phase was developed from a preliminary questionnaire produced in the autumn of 1973 for the Swedish Confederation of Salaried Employees (TCO) (*The AFTA Questionnaire*, Bradley, Börjeson, and Lundgren, 1973). The qualitative analysis was used, inter alia, to modify and develop the questionnaire. It was thus adapted to the problems that were covered in this project. It contains over 350 separate questions, primarily with fixed alternative answers. The fourteen areas in the questionnaire correspond by and large with the areas in the theoretical model in Figure 2.

The questionnaire's design was based on two methodological considerations. Firstly, we wanted to study the impact of computerization on the psychosocial work environment. Here, we wanted to compare the response patterns for different groups of employees within the organization that differed in significant respects in terms of their use of EDP. Secondly, we wanted to use a questionnaire where the subjects were asked direct questions about the impact of computerization (the EDP system) on different conditions at work. These 'computer effect questions' are in the instrument as supplementary questions after the 'direct questions'. We often structured the 'computer questions' in a similar way to the other questions in the same part of the instrument. This allows the instrument to be used in follow-up studies and different types of cross-sectional studies.

The method for analyzing data included a qualitative analysis of the interview material, descriptive statistics and certain multivariate analytical methods.

The group investigated in the quantitative phase was formed by a stratified random sample of the approximately 5000 employees in the organization. Six departments with some major differences in terms of their connection with EDP technology formed sub-populations.

3. Summary of results

The following changes in the design/structure of the organization were directly attributable to the introduction of the computer system:

- a decrease in the number of levels in the hierarchy, due to the disappearance of some positions;
- a special key-punching centre was closed down;
- the Data Processing Department was enlarged;
- a special department for locating/correcting errors arising from data handling was created.

The changes resulting from the introduction of the EDP system mainly affected the staff in the Data Processing and Operating Departments (the bookkeeping departments).

The operations departments' staff have a shorter training and lower qualification requirements than staff in the Data Processing Department. Reasons for discontent are found in both divisions, but these are different and have different effects in terms of health, for example.

Some 2000 people work in the *Operating Departments*, but the number drops or is expected to drop for every step in the automation process (see Chapter 1, section 1). There is dissatisfaction in this department, stemming from the removal of the more interesting and stimulating tasks. Only a few work tasks remain, and the personnel feel deprived of the responsibility that used to be a part of their jobs. They feel insecure, they are afraid that 'the computer is taking over' and they are worried about being relocated. The degree of interchangeability is high; it is easy for these employees, with their broad-based skills, to transfer to other departments, and they can easily be replaced. These employees have less influence in planning their work. They often have to contend with overtime and unscheduled working hours because of down time in the computer room, and this makes it difficult for them to plan their leisure time and family life. The pace of work is both intensive and machine-controlled, and the personnel feel that the pace is too hard. Daily deadlines are a feature of the work schedule, and this serves to increase the workload in all departments. The growing amount of strain in the work has made inroads into what was a high level of staff morale and a good working atmosphere.

Certain physio-ergonomic problems intensified with the introduction and operation of EDP technology. The findings show major physio-ergonomic problems at a traditional white-collar workplace.

The number of employees in the *Data Processing Department* is growing. These employees feel a greater sense of security, since people with computer skills are still in short supply. However, the Data Processing Department is not markedly different from other departments in terms of employees' attitude to work content. But these employees find it difficult to relax and leave their jobs behind them after work, and often go to work even when they are ill. They feel some psychological strain, caused partly by the fact that there is no-one to stand in for them at work. The amount of influence on work has also decreased for staff in the Data Processing Department, particularly in terms of the pace of work and choice of work methods.

The introduction of the computer system placed some psychological strain on data processing personnel, one reason being their fear of errors in their work, producing anxiety and worry. The consequences of error in these jobs may be far-reaching and may affect both the company's Operating Departments and the customers. Data Processing Department personnel claim that too much responsibility has been placed on their shoulders.

The Data Processing Department has become isolated within the organization, separated from other departments by an 'iron curtain'. The Data Processing Department can also be described as the organization's 'buffer zone'. It appears that the frequently contradictory demands from various quarters are concentrated or accentuated here.

Computerization has meant that people are given work tasks/positions that can be put under one or more of four headings: 'too much', 'too difficult', 'too little' or 'too easy'. This situation gives rise to stress reactions and the organization becomes divided into clearly defined departments which, among other things, has an effect on personal relationships between employees.

From a basic organizational perspective, it may be said that the Operating Departments have a number of work environmental problems that may be attributed to understimulation (both qualitative and quantitative) while the Data Processing Department in many respects shows signs of problems related to overstimulation (see also Chapter 7).

The structure of the quantitative instrument, in which approximately 350 questions were shown to measure forty main factors, is presented in a separate research report (RAM report no. 6). The factors relate to theoretically anticipated dimensions. Indices were constructed which subsequently formed the basis of secondary analyses (index = scales or groups of questions, to improve the reliability of the findings). The indices measure the following factors:

- the general attitude to work
- the psychosocial work environment
- industrial democracy and influence at work
- the effects of the data processing system on work environment

- physical and psychosocial symptoms
- the content of leisure time
- the effects of work on leisure time/family life.

All indices were investigated on the basis of different individually-related background data (e.g., age, family status, whether the individual had children and, if so, their ages and educational background) and work-related background data (e.g., position, income, period of employment, the scope and character of working hours, contact at work with the general public). The findings were reported in diagrams with numerical indications of significant differences.

RAM report no. 6 also contains a section describing different homogeneous sub-groups in the material examined. For example, on the basis of different indices (see above), we analyzed which fundamental values related to work and work/private life can be identified (latent profile analysis). In addition, we applied a separate method (regression analyses) to answer questions such as: Which areas in the socio-psychological environment best 'explain':

- general job satisfaction
- nervous complaints
- psychological stress
- short-term absenteeism
- the number of days of absenteeism
- the effects of work on leisure time/family life?

The following brief summary of results is taken from RAM report no. 6 .

4. *The variables in the study*

Figure 3 summarizes the variables used in the analyses, including indices and background variables. The indices are related to the project's original theoretical model (see Chapter 3, Figures 1 and 2). The following groupings of variables are to be found in Figure 3:

- work-related background variables
- individual-related background variables
- weighting of work environment factors and general relationship to work (indices)
- leisure time activities (indices)
- work environment conditions (indices)

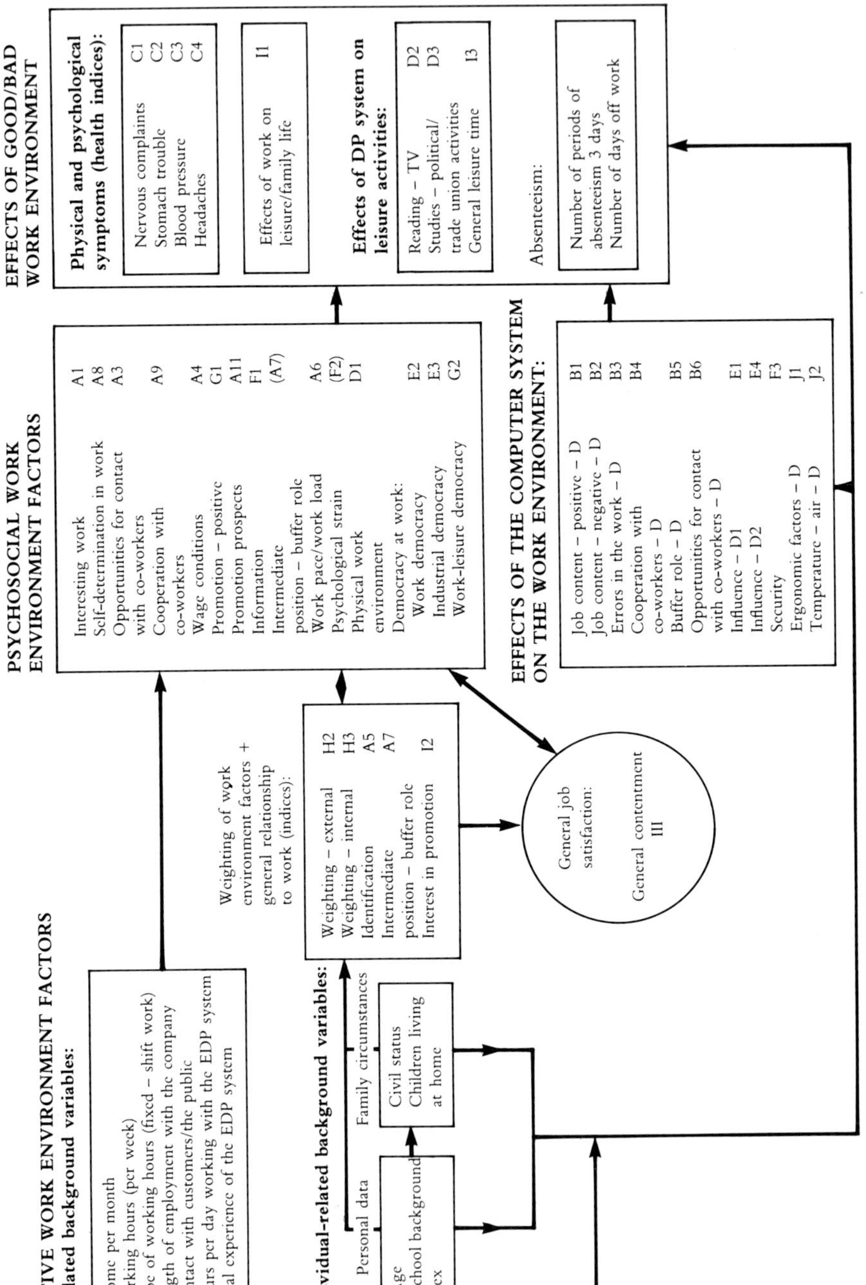

Figure 3. A summary of variables and indices (Bradley's models)

- the impact of the computer system on the work environment (indices)
- physical and psychological symptoms (indices) and absenteeism.

This model is intended to provide an overview of the investigation variables in the empirical material and to help in analysis. The arrows illustrate the hypotheses that are the basis of the analyses described below.

5. Additional problems

The relevant questions are as follows:

1. What areas of the psychosocial work environment help most to 'explain'
   - general job satisfaction (general contentment) (H1)
   - nervous complaints (C1)
   - psychological strain (F2)
   - the effects of work on leisure time/family life? (J1)

In these four analyses, fifteen different indices (see below) are used as independent variables to illustrate the psychosocial aspects of the work environment (in box in Figure 3) as follows:

Interesting work (A1)
Opportunities for contact with co-workers (A3)
Pay conditions (A4)
Identification (A5)
Work load (A6)
Intermediate position/buffer role (A7)
Job autonomy (A8)
Cooperation (A9)
Promotion prospects (A11)
Physical work environment (D1)
Work democracy (E2)
Industrial democracy (E3)
Information (F1)
Promotion, positive (G1)
The effects of work on leisure time/family life (I1)

2. What effects of the computer system (D-indices) make the greatest contribution towards explaining:
   - general job satisfaction
   - nervous complaints
   - psychological strain
   - the effect of work on leisure time/family life?

In these four analyses, eleven separate indices concerning the effects of the DP system (D-indices) are regarded as independent variables (in box in Figure 3). The same indices/variables as in the first set of regression analyses recur here as dependent variables. The independent variables (D-indices) are as follows:

Job content, positive—D(B1)
Job content, negative—D(B2)
Errors in the work—D(B3)
Cooperation—D(B4)
Buffer role—D(B5)
Opportunities for contact with co-workers—D(B6)
Influence—D1(E1)
Influence—D2(E4)
Security—D(F3)
Ergonomic factors—D(J1)
Temperature—air—D(J2).

Thus the aim of the analyses presented here was to study which areas in the psychosocial work environment did most to explain different criteria of satisfaction at work. In a similar way, the different effects of the computer system, which were reported by the people interviewed, were used as predictors of psychological strain at work. Multiple regression analysis was used which, together with the individual questions in the index measurements, is described in greater detail in the RAM report no. 6.

The problems are related to the model of investigation variables in Figure 3. Both the different indices that measure various work environment factors and the different indices that measure the effect of the DP system on the work environment are regarded as independent variables. Variables that are considered dependent are to be found mainly to the far right of the model, i.e., nervous complaints, the effect of work on leisure/family life, different measurements of absenteeism. One of the work environment indices, namely psychological strain, is considered as a dependent variable in the regression analyses. In addition, general satisfaction has been included as a dependent variable. These indices are in the circle in Figure 3 to allow a theoretical connection to the background variables and 'weighting variables'.

6. *Findings: Analyses of the psychosocial work environment*

- What factors in the psychosocial work environment helped most to explain *general job satisfaction?*

It is primarily the companionship and cooperation with co-workers (A9) and the character of the work in terms of interesting content (A1), which explain satisfaction at work (H1). According to the analyses, contributing factors also include identification with the formal company/organization (A5), personal development and promotion prospects at work (A11) and the degree of job autonomy (A8). The physical work environment (D1), which makes a fairly significant contribution towards explaining job satisfaction, should also be considered—in this case, in a traditional office environment. However, earlier findings reveal a series of physio-ergonomic problems at this workplace which are often related to the mechanical equipment (RAM report no. 2). Somewhat inexplicably, according to the regression analyses, the two democracy indices have not contributed at all towards explaining job satisfaction, E2, or E3. Separate analyses show that on the basis of these independent variables it was possible to predicate the general job satisfaction fairly well (59 per cent of the total variance was explained by the independent variables).

The findings supported the traditional alienation theory views of work environment. These theories emphasize the importance in terms of job satisfaction of the following needs: the need for interesting, meaningful work with scope for self-determination (job autonomy), the need to identify with a work group, the opportunity for contact with other people at work and the opportunity for normative integration at work (see Chapter 3). The factor of cooperation with co-workers (A9) is of particular interest as it is the clearest explanation of general job satisfaction. This implies that personal relationships at the workplace are a central and important component of the work environment. More details of the factors that contribute to, and prevent, good personal contacts are to be found in *The Working Roles of Secretaries and Word Processing Systems* (Bradley, Bergström and Sundberg, 1984); see also Chapter 6.

- What factors in the psychosocial work environment contributed most towards explaining *nervous complaints*?

The separate factors in the work environment provide little to explain the dependent variable, nervous complaints. (Only 12 per cent of the total variance in the dependent variable can be explained in terms of the independent variables.) However, work place and work load (A6) help explain the occurrence of nervous complaints. If the pace of work is too high, too irregular and too dependent on machines or other people, and if the individual feels that the work is hectic and stressed, then nervous complaints tend to appear.

- What factors in the psychosocial work environment contributed most to explaining *psychological strain?*

A relatively good prediction or explanation of psychological strain at work can be made on the basis of the different factors present in the work environment. (38 per cent of the total variance in the dependent variable can be explained in terms of the independent variables.) Psychological strain can be explained primarily by workload and work pace (A6), in other words, if the pace of work is too hectic, too uneven and controlled by machines and other people to too great an extent, and if the work is perceived as hectic and stressful, there is a tendency to feel psychological strain. Even the intermediate position and the buffer role (A7), job autonomy (A8) and interesting work (A1) are fairly prominent factors in explaining psychological strain. The findings show that the more interesting the work is perceived to be, the greater the psychological strain it involves. There are no corresponding findings for autonomous work. Here, a high degree of strain at work is more easily related to a lack of job autonomy.

- What factors in the psychosocial work environment contributed most to explaining the *effects of work on leisure time/family life?*

The psychosocial work environment variables explain 26 per cent of the total variance in the effect of work on leisure time/family life. This indicates that the various factors in the work environment go some way towards explaining the effect of work on family life and leisure time. It is primarily work pace and work load (A6) that explain this dependent variable. Whether or not the employee is satisfied with his salary (A4) and his perception of the prospects for promotion and personal development in work also have some effect (G1).

7. *Findings: Analyses of effects of computerization*

- What effects of the computer system (D-indices) contributed most to explaining *general job satisfaction*?

According to the regression analysis, only an insignificant part of the variance in the dependent variable of general job satisfaction is explained by the independent variables concerning the different effects of the computer system on the work environment. This may be due partly to the fact that the arguments become rather less direct: what effect does the computer system have on general contentment at work? However, it may be interesting to examine what DP effect indices contribute most to explaining general contentment. The index that measures the impact of the computer system on different ergonomic conditions at work (J1) is the most prominent. The impact of the computer system on job content, represented by two indices (B1 and B2), and the computer system's impact on the opportunities for personal contacts at work (B6) also occur as explanatory variables.

- What effects of the computer system have contributed most to explaining *nervous complaints*?

Here too, a relatively insignificant part of the variance in the dependent variable, nervous complaints, is explained by the independent variables. The explanation of the occurrence of nervous complaints is spread fairly evenly over four indices: the effects of the computer system itself on job content (B2), the computer system's effect on the risk of errors in work (B3), the computer system's effects on opportunities for contact with co-workers (B6) and the effects of temperature and air conditions at the workplace (J2).

- What effects of the computer system contributed most to explaining *psychological strain*?

A relatively insignificant part of the total variance in the dependent variable, psychological strain, is explained by the effects of the DP system. Three indices contribute to more or less the same extent towards explaining the occurrence of psychological strain: the computer system's effect on the buffer role (B5), the computer system's effect on the risk of errors in work (B3), the effects of various ergonomic conditions (J1). The computer system's effect on job content (B2) also occurs here, albeit with rather less significance.

- What effects of the computer system contributed most to explaining *the effects of work on leisure time/family life?*

This question may be expressed more simply: does the computer system indirectly affect family life and leisure time? First, it may be stated that the variance in the perceived conflict between work and conditions of life outside work is explained by the effects of the computer system on different conditions in the work environment. The effects of the computer system on ergonomic conditions in work (J1) contribute most to explaining the perceived conflict between work and leisure time/family life. The impact of the computer system on job content is also present as a factor that explains whether one feels that work impinges upon one's leisure time and family life (B2).

RAM report no. 6 gives a more detailed description of these results, with the material presented in tables.

8. Action taken

The findings of the different phases of the projects were discussed in different groups within the Postal Giro Service and with the joint local reference group. A separate chapter in sub-report 2 (Bradley and Nilsson, 1976) was devoted to the action taken at different levels. One year later, a summary report (RAM report no. 4) was presented which dealt further with the proposed measures. The following section is an excerpt from Chapter 6 of RAM report no. 4.

Measures that could be applied generally to work environment problems in the broad sense were divided into measures at the different levels discussed in relation to the concept of the work environment, i.e., the community, the organization (company, trade union), the individual. We maintained that measures in the area of the work environment/computers should also be tackled at these different levels. These general measures are discussed in a separate chapter. They were subsequently debated at the organization that was the subject of the study. The discussions carried on in conjunction with the evaluation of the results that emerged in the project took place both in the local reference groups and in the company's personnel and environmental committee, which consisted of representatives from the trade union, management, the company health service and staff consultants.

The Data Processing Department and the bookkeeping departments had more problematic work situations than the other departments studied. The Data Processing Department was clearly in an intermediate position (role ambiguity) in the company and demands were placed upon it both from other departments within the company and from the general public. Any error occurring in the Data Processing Department had immediate repercussions on both the personnel in the bookkeeping departments, who had to suffer long waiting times, and on the general public. Time margins were very tight because all cases were supposed to be processed the same day. The computer staff expressed their dissatisfaction in a number of areas, for example, work loads, physio-ergonomic conditions, promotion, working hours. In the bookkeeping department, the dissatisfaction appeared to be mainly related to the content of the tasks performed: it was considered that the work contained too few tasks, the staff generally saw less point in their work, the work did not provide the opportunity to make use of individual resources, and it was felt to be uninteresting. Computerization had served to aggravate this trend. In addition, physio-ergonomic problems were reported: there was considerable discomfort from noise and work positions at the mechanical equipment.

In this company, the subject of the study was primarily a batch-type system. Batch processing involved alternating manual and computerized processing of material. Some result patterns were directly related to this factor. Technological trends were towards a greater use of online systems using terminals for both entry and retrieval of information, involving new opportunities and new risks (see Chapter 5) for work and working conditions.

Prior to the introduction of the computer system in the company, specialization had often taken place which had long been considered as giving greater 'efficiency'—a philosophy rooted in 'Taylorism'. Taylor's system involved using measurements and analysis to check work methods and time. An effort was made to apply the advantages of large-scale production to office work. Office work became a 'stationary job': in order to save time, everything needed should be readily available within reach of the worker. As was the case when new technology was introduced on the workshop floor, the problems of job allocation appeared first, followed by the problems of mechanization—two problems that are closely related. A strict allocation of work tasks among various groups in the workforce, coupled with the effects of wage policies and other measures, paved the way for the automation of work processes. The entire work flow in the office was divided into many small operations. Some of the tasks that had belonged to office work were now transferred to computers, and specialization became even more pronounced (however, the organization's goals have been broadened in recent years, which is underscored by the new labour legislation, among other things, see Appendix 6).

It is important that organizational changes and changes in jobs resulting from the introduction of computer technology be discussed at the planning stage of the process of computerization. If a major part of an employee's work is to be computerized, alternative work tasks and the concomitant training should be discussed at an early stage and/or other division of labour principles should be examined.

Problems occurring at one level are often a reflection of events at other levels, and in the final analysis they reflect the social structure. The fact that, for competitive reasons, the company that was the subject of this survey ran a twenty-four-hour service, i.e., all incoming cases were to be processed the same day, is related to the fact that our social structure contains a number of competitive aspects. It was our contention that there was a risk of overemphasizing the competitive factor and we suggested a survey to provide a more detailed picture of the general public's/customer's attitudes and service requirements. In our opinion, an obvious goal would be to avoid having to run a twenty-four-hour service. It was impossible to reconcile the technical capacity and type of technology available (batch processing) and the current levels of service with a good work environment.

Fears of making mistakes in one's work emerged as being an important cause of psychological pressure in the Data Processing Department; the risk of making errors had increased, as had the consequences of errors. Delays in work, either because of manual errors in the computer room or because of breakdowns, had immediate effects on working hours, sometimes necessitating overtime, mainly in the bookkeeping departments. This work environment problem was largely caused and aggravated by the demand for twenty-four-hour service.

Since the findings indicated tension between the Data Processing Department and the bookkeeping departments, it was important to increase understanding for, and knowledge of, the work conditions that prevailed in each department in the organization by, for example, increasing the amount of information/training given. Thus it was proposed that the bookkeeping departments be given training in the field of data processing, both in general computer technology and on the actual work situation in the Data Processing Department. A course in computerization/work environment and an opportunity for short periods of first-hand experience of working in the Accounts Department were proposed for Data Processing Department staff. We also emphasized a general need for more training in organization and work environment issues for the computer specialists (systems staff and programmers), which should include discussions on problems related to subjective values.

In an attempt to reduce the prevailing communication problems, efforts should be made to create a 'data language' that is more easily accessible and can be used by larger groups of people, not only by 'professional people'.

The findings indicated that departments within the company were isolated from one another. The staff from one department knew little about the work of other departments, and this reduced the individual employee's opportunities to place his or her own work situation in a wider perspective, a factor that often helped create alienation. We proposed a presentation of the work area of each department, for example, by means of articles on each workplace in the staff journal, and we also proposed that the trade union publication should be responsible for the information aspects of this problem. Some steps have been taken in this area.

According to our findings, the scheduling of working hours was a factor in the work environment that contributed to the occurrence of psychosomatic complaints. This applied both to the bookkeeping departments and to the Data Processing Department. The working hours were indirectly related to the computer system. In our opinion, it should be possible to make phased improvements to the computer equipment.

We also felt that efforts should be made to abolish the three-shift system in the Data Processing Department. Working towards this goal also involved demands in areas that were related to the twenty-four-hour service. It was our opinion that the negative effects on the individual's health and family/private life were sufficient reason to abolish the three-shift work schedule. Moreover, unless the three-shift system were changed, the Data Processing Department would encounter considerable difficulties in recruiting new staff. In the bookkeeping departments, working hours should be arranged to allow for improved and more reliable planning for the personnel's 'work-free' time.

A large number of employees were exposed to physio-ergonomic problems such as sound/noise conditions, work positions, the mechanical equipment and the way the work premises had been modified to suit the work itself. Athough these problems would be solved by moving to new premises (planned for 1979), it was important that immediate improvements should be made.

The psychological stress that existed in the Data Processing Department was partly related to the high degree of specialization and the fact that no replacements could be found if a staff member was off sick, for example. We recommended an internal investigation to assess different ways of relieving this pressure. Data Processing Department personnel themselves had reported that the department was undermanned.

In concluding, we emphasized the fundamental importance from the work environment viewpoint of an assessment of the volume of work and work resources. An important step in the planning process is to follow, as far as possible, the trends in production volume over a longer period in order to forecast changes in work volume and the resources available to perform this work. In both the bookkeeping and Data Processing Departments, the work load was considered to be too heavy. We concluded the chapter on action to be taken as follows:

> The company is on the threshold of yet another phase of automation. The investigation reported here has considered different areas of the work environment and examined the impact that computerization has had on them. It is hoped that the

measurement instruments used in this process may be of benefit to users in different phases of a process of change, i.e. in evaluating not only a system in use but also planned EDP systems. It is vital that the goal of a good environment is incorporated into systems development work in respect of a number of aspects of the work environment: employee influence, job content, organizational design, contact, information, etc.

Various measures were introduced in the organization at different stages of the project, related both to general work environment problems and to problems more directly linked to computerization and the computer system. The following measures were either planned or implemented.

- A review of the relationship between customer demands and company operations was begun.
- A general review of data processing was carried out at the company in cooperation with external technical expertise. A number of measures were proposed, including improvements to the technical system.
- The introduction of computer technology had accentuated some physio-ergonomic problems, primarily in the bookkeeping departments. Improvements were introduced in stages. Particular attention was focused on the coding machines (their work position and noise levels). An external research group was to examine further the ergonomic problems.
- Training courses in computer technology were started for personnel from the Data Processing Department, and work material was produced.
- Descriptive articles were written for the company's staff journal, featuring different departments, these articles including interviews, illustrations and the like.
- We maintained that various alternatives should be examined for the people whose jobs have been depleted in different ways as a consequence of the introduction of computer technology in order to enrich their job content, for instance, by introducing various kinds of job rotation and job expansion schemes. This work would not be an obstacle to the introduction of preventive measures. Discussions were begun on measures related to work organization and the division of labour in the operating departments.

A special 'theme day' with some hundred specially invited people, who in some way were affected by or were involved in work on new technology-working life issues, dealt with the final results of the report and the measures proposed by the project. This was carried out both in groups and in plenary sessions.

This was followed by feedback of experiences from different organizations and companies, in order to give the research findings a wider circulation outside the organization. Other companies and government bodies also found themselves facing a major process of reorganization related to computers. A series of articles and books was produced (see Appendix 2). The members of the project attended a large number of external conferences, courses, etc., so that their experiences would reach a wider group. (At that time there were far fewer broad-based studies of the psychological and social impact of computerization.)

A book, '*Computer Technology, Working Life and Communication*' (Bradley, 1977), was commissioned by the delegation for long-term research, in which the preliminary results from the Postal Giro project were published. Among other things, the book contained an analysis of rationalization, technical change and computerization. It dealt with computer technology and its effects on employment and occupational structures, and one chapter was devoted to the main areas for future research. Some 6000 copies of the book were circulated under the auspices of the Research Council, the circulation list including politicians, research institutes, institutes of higher education and the national employers' and employees' organizations. This book may be said to have helped to shape opinion and focus attention on areas concerning computerization at work and in the community. One year later, two government surveys were started to examine this problem area.

Six trade union consultants employed at the Postal Giro Service were involved with the processes of change that occurred with computerization. This involvement became one of the largest projects run by the Swedish Confederation of Salaried Employees (TCO).

Since then, further work at the Postal Giro Centre has been published in '*The Postal Giro in Stockholm*' (Grip and Sundström, 1984). The book focuses on the work done by the trade union and is a detailed description of the various stages in the process of employee influence, and analyzes the position of the trade union. On the subject of employee influence, the authors state that as a result of recent years' trade union work, management has become more willing to listen to the opinions put forward by the employees and the relationship between the actors on the labour market as clarified at the Postal Giro service.

It is considered that the co-determination agreement covering government employees (MBA-S) and the regulation by agreement of projects' terms of reference have given the trade union greater influence on work organization and production methods. Grip and Sundström also state that at the time (some ten years after the RAM Project completed its investigations at the Postal Giro Centre) work organization was still a cause of lack of contact between employees and made cooperation difficult. The book also claims that influence at existing workplaces, with extremely high muscular and skeletal loading, is limited; it is still difficult to influence the scheduling of working hours, which are an integrated part of the production system.

How, then, has greater influence on work organization and production methods been used? In comparison with previous restructuring processes, the authors say that project work is better handled in the local collective agreement on the development of the Postal Giro Centre's new production system (PGP). Influence has been won in issues related to staff shortage and undermanning, and influence can now be brought to bear on appointments and the recruitment of replacements.

A new way of assessing the operation was presented in the local postal giro collective agreement and the concepts of efficiency contained in the co-determination agreement for government employees: changes must produce better jobs, jobs must not be eliminated and customers must be satisfied. However, it is difficult to say with any certainty whether these changes in operational criteria have taken root. Grip and Sundström state that employers must develop methods that demonstrate that they have met the MBA-S and PGP agreements' demands for a better work environment and better work organization, and have maintained employment levels and the same standard of customer service.

Against the background of experience gained from the RAM Project and documented experience from the work of trade union consultants, the conditions now exist for making these measurements. Thus there is scope for a third research project to follow up and measure the effects of the consultancy work and widespread trade union activity on employees' working conditions and their psychosocial work environment. In the autumn of 1985, the Postal Giro Service's new production system was to be started. The Postal Giro is still one of the country's most computerized office workplaces. Most of the workforce, currently 5500 strong, are women.

The objective of maintaining the competitive position of the Postal Giro Centre vis-à-vis other payment systems remains. In their concluding chapter, Grip and Sundström return to the concept of competition:

> A political mandate makes the Post Office Administration and the Postal Giro Centre the organizations used by the Government for its payments. The management of the Postal Giro Centre

sometimes refer to their operation as if it were a private company which was exposed to severe competition. If the Postal Giro Centre is incapable of maintaining its competitive position, according to its management, this would mean the Postal Giro Centre would not survive. We have been unable to find anything to support this view. In addition, it is strongly misleading to make the biological analogy that an organization is a living thing that will—preferably—outlive other organizations. Government operations can be organized in a number of ways. Restructuring this operation is not a life-or-death matter. It is unclear why management puts forward these dramatic descriptions, particularly when this kind of statement is not to be heard from a higher level. It is possible that the metaphor of competition is used as a simplified description of the operation. Instead of developing measurements of the operation per se, they prefer to sub-optimize and place their operation in relation to corresponding operations that are not in the public sector. (Grip and Sundström, 1984, p. 170)

5.
Experiences from Company 2: The Trygg-Hansa Insurance Company

The project's second field survey took place in a company that provides a service—Trygg-Hansa, an insurance company. We concentrated on the company's head office, which is located in the Stockholm area and where approximately 1000 people are employed. This company has an online data processing system, in which the use of video display terminals was an important and innovative application. It was considered important to chart in detail changes that took place in the work and working conditions when these terminals were introduced. It was still possible to include a comparison group, a group of employees in the company that did not use terminals, but worked under the same conditions (see the follow-up interviews at the end of this chapter).

1. Computerization at Trygg-Hansa

The following description is based on the data that were available when the survey was carried out and the report written, as well as some current information from 1985. Trygg-Hansa has many years' experience of the use of office machinery and equipment. Over the last decade or two, computers were introduced for batch processing. An insurance register was compiled and stored in memory in 1972, and was made available to the insurance company offices in microfiche form. A trial project using direct-linked VDTs was started in 1974.

Background

The terminal systems studied in the RAM Project were based on 'System Plan 72', which involved a major investment in computers. A two-stage trial project on VDTs was started in 1974, covering departments such as Motor Claims, Motor Insurance and Customer Service. On the basis of the experience gained in this trial project, the company's Board approved the installation of 230 terminals and ancillary computer equipment, to be brought into use in early 1977. Technical specifications were defined in collaboration with consultants, and a reference group was appointed consisting of people from both the head office and the field organizations. It took five months to evaluate the tenders, and the terminal supplier was selected on the basis of the price of the terminals and the extent to which they met ergonomic standards.

These 230 terminals were to be installed both at the head office and at the claims offices and were intended for use in personal and property insurance transactions. The aim was to increase productivity and improve customer service. It was estimated that the work force would be cut by 100 people in 1980 and that the system would pay for itself within five years.

A description of the technology

The terminal system consisted of video display units. Printers could be connected to the terminals so that users could make a printout of the information on the screens. The terminals were connected to a central computer at the head office by the local network or by permanently installed cables. The system would normally be available during office hours, with a 1986 target of 96 per cent availability.

The software system was of the dialogue type. A relatively large number of different types of case had to be processed in the various operational systems. Each type had one or more dialogue routines. These involved entering data from the display units, which were immediately verified and processed in the central computer, and the display on screen of data stored in databases and data entered and processed. A normal case involved between two and eight separate screen displays. Only a limited number of terms and a limited number of instructions could be entered on each screen. The screens were divided into fields each of which had a clearly defined function and design. Different types of case were processed by selecting routine options from a standard display menu which was shown at the beginning and at the end of the dialogue routines. Finally, terminal users had access to a special help function for all the screen displays. Every user had to identify himself by means of a personal code on a magnetized card. This code determined the user's degree of authorization, i.e., what information he or she had access to and could alter.

Microfiche continued to be used in the sales offices, but plans were made in 1977 to start a new trial project to examine the possibility of installing terminals at the sales offices as well. In 1980, terminal-based query and data entry systems came into use in the Property Insurance Department and query systems were installed in the Personal Insurance Department, which meant that, with the exception of personal insurance, virtually all the mechanized office procedures for processing insurance policies were carried out on a terminal system.

Terminalization and work methods

The company decided that the terminal would be the principal aid for the individual user. Changes were made in three main work areas: retrieving information on insurance policies, updating insurance policies, and making routine calculations.

Formerly, one of the following methods was used for searching for insurance policy data: a manual search through data lists, through the microfiche register and through manually produced registers, or by telephoning the head office with computer queries which, after being processed in the central computer, were answered one day later. Now, all information is displayed at the terminals. The aim is to be able to display on the terminal screen all insurance data that may be needed to process a given case. (The terminal also accesses other registers, for example, the National Vehicle Register, which removes the need to telephone this severely overloaded source of information.)

The basic method for updating insurance data in the old system was to register data on one of several kinds of form. These forms were then coded and sent to the Coding Department. After the central data register had been updated, error lists were produced which were returned for correcting. In some areas processing, registering, printout of policies, etc. had been done manually. Updating is now increasingly done on the terminal and when data are entered, a number of error checks are carried out on the data being displayed and processed, allowing corrections to be made on-screen. Routine calculations based on customer data, e.g., the size of premium, are now done on an online computer terminal.

Systems development plans

What was the background to the next stage of systems development planning? In common with other large insurance companies, Trygg-Hansa had introduced punched card technology and EDP at an early stage in order to improve the efficiency of insurance policy processing in the company. The traditional areas where EDP was used in the insurance sector were: keeping the large property and life insurance registers up-to-date; issuing notifications of premiums due; paying of pensions, annuities and claims; processing statistics, etc. In the 1950s and 1960s major savings were made in these operations by introducing punch-card technology and EDP. This meant that a central data register had to be created, which had to be accessed by means of lists etc. Because of the inherent disadvantages of the long updating times and out-of-date information contained in the lists produced for the batch system, the staff willingly accepted the new direct memory access, communications lines and terminal equipment which allowed information to be transferred and displayed on-screen in seconds.

As mentioned above, Trygg-Hansa began to develop guidelines for introducing a new terminal system in 1972. These guidelines and plans for the applications to be developed were called Systems Plan 72. One of the objectives of this systems plan was to create EDP systems that were as independent as possible of the various divisions within the insurance company. This involved producing general claims systems for certain sectors such as home, house, boat and motor insurance, general premium processing systems, etc.

In an attempt to make the work of the application programmers easier, a start was made in 1974 on producing comprehensive systems programming software which handled communication between the database processing system, the data communication systems, etc. and the programmers. This programme included a data catalogue function and a programme generator (generating standard programmes for checks, etc.). These new help systems were first used in an internal training package for users.

At the time of the survey, the development of Trygg-Hansa's own advanced systems programme software had run into some snags which delayed the development of the application systems. Problems were also encountered in developing a completely general EDP system that was not specific to any area of the company's insurance business. The Trygg-Hansa customer policy account system was one of the terminal systems introduced in the spring of 1977, and is an example of a general system. The customer's premiums for different types of policy were totalled, the payments were spread over the year and the customer policy account system transferred money each month from the policyholder's bank account to Trygg-Hansa. The customer account system can be compared to the personal current ledger routine. The service it offered the policyholder attracted a rapidly growing number of customers to this payments system. The customer payments system is run by the head office, and queries may be sent to the head office from local branches that are equipped with terminals.

Essential customer information is extracted from Trygg-Hansa's personal insurance register and stored in a database. The customer service section at the head office and at the local branches that are equipped with terminals use this information in customer contacts, sales, etc. The system came online in 1977.

In the autumn of 1978 a new EDP system was started in the

Company Insurance and Industrial Insurance sections, from where information is also keyed into the system. The system aims to reduce transactions times, cut costs and allow the automatic indexing of premiums to compensate for reductions in the value of money, etc.

In the first phase some forty terminals were installed at the head office. More terminals were installed at the head office in the second and third phases, which were started in 1979 and 1980. The local branch offices can access the register with queries. When the EDP system for company and industrial insurance was started, approximately eighty terminals were installed at the head office and at local branch offices. All routines use a dialogue system on the terminals. This means that several screen displays containing different processing options are used for the same transaction. In January 1980 approximately forty terminals were installed to process vehicle damage claims from about ten claims offices throughout the country. At the same time a trial project was started for processing vehicle insurance on VDTs. By 1980 a total of approximately 300 terminals were in use. At that time the company had 3300 employees.

Table 5.1 shows the development of hardware capacity in recent years at Trygg-Hansa. In addition to the number of VDTs, details of hardware capacity are included. The number of terminals has increased from approximately 100 to approximately 2000 since the RAM Project concluded its work at the company (1978/79). Today, there is an average of one and a half employees per terminal. There are also about 100 printers and PCs that total twenty-five MIPS, i.e. the same as the CPUs in Table 5.1. The figures illustrate an initial total investment in mainframe computers followed by the introduction of

Table 5.1 Development of hardware capacity of Trygg-Hansa, 1978–1986

		Performance measurements			
Time	CPU types	MIPS	MB	GB	VDTs
78–01		1.4	4	10	ca 60
79–02	IBM 148 + 158	3.2	6	18	ca 150
80–10	IBM 148 + 3032	6.9	12	22	ca 300
82–05	IBM 3033N + 3032	9.3	16	25	ca 900
83–05	IBM 3033N + 3033U	9.3	24	34	ca 1200
84–04	IBM 3033N + 3033U	19.5	48	48	ca 1500
86–01	IBM 3081K + 3033U and 24 IBM 8140s	28.0	64	55	ca 2000

Abbreviations:
MIPS = CPU speed in millions of instructions per second.
MB = primary memory in CPU Mega (million) bytes.
GB = disc memory in Giga (billion) bytes.
CPU = Central Processing Unit.
Source: Aktuellt från Systemtjänst, 4 (1985).

minicomputers and even small computers using microprocessors, for example, personal computers.

Terminalization and organizational change

The main changes in organizational structure when VDTs were introduced may be summarized as follows:

- a reduction in the number of levels in the organization; as a result of a number of mergers, there were too many supervisors at the intermediate level;
- an expansion of the Data Processing Department, in particular the commissioning of a large number of consultants from outside the organization;
- a special word processing department was set up;
- even before the introduction of VDTs, the company had democratized its structure, by introducing employee representatives into two management groups at different levels in the organization, and this made some of the changes easier; thus, a more appropriate organizational structure had been established before the technology was introduced;
- the simultaneous introduction of modern open-plan office design and display terminals caused some conflict between these two methods of rationalization, for example, because display terminals had special office lighting requirements which were hard to reconcile with many open-plan office layouts, the display terminal work stations had to be screened off.

2. *Research design and methods*

An online computer system was in use at Trygg-Hansa. The personnel had direct contact with the computer from their VDTs. We collected data by means of interviews, first-hand observation, a study of the company's internal documentation and questionnaires. The general approach was in the nature of a follow-up survey. The process of introducing VDTs was observed over a two-year period and the work was carried out in the following sequence:

1. Quantitative measurements before the introduction of the data processing system, involving all personnel at the company's head office, with the aim of assessing employees' expectations and fears about the introduction of the new VDT system (December 1976);

2. A semi-structured interview with the users fo. a query-type terminal system (December 1976); the objective was problem identification;
3. Semi-structured interviews with personnel who had completed training and had just begun work at a new dialogue-type terminal system (May 1977).

The original plans had to be modified when the company's computerization programme was halted in the autumn of 1977.

4. Interviews with personnel who did not use their VDTs even though they had completed training and had undergone various forms of preparation for VDT work (spring 1978);
5. Interviews on the modifications to the rationalization plan with experts who represented the trade union, the company, the administrative rationalization personnel and personnel from the Data Processing Department; the objective was to illustrate general trends (spring 1978);
6. Quantitative follow-up measurements throughout the head office, after the new terminal system had been introduced; we approached all personnel who were directly or indirectly involved (late 1978, early 1979).

The questionnaire survey completed at the end of 1978 and the beginning of 1979 covered the same population as the quantitative measurements of December 1976, before the data processing system was introduced.

The work was carried out in close cooperation with a local reference group consisting of two personnel representatives, the personnel head and the head of systems analysis. Personnel representatives from the departments directly involved in the research group's fieldwork were seconded to the reference group, as were a physiotherapist and company doctor.

Experiences from the different phases of the fieldwork at the company provided different kinds of knowledge about computerization. The results of the preparatory measurements will be described here: the interviews carried out in 1976 (points 1 and 3 above), and the quantitative follow-up measurements that were done in 1978/1979 (point 6 above). The questionnaire survey was based on a random sample of approximately 1000 employees at the company's head office.

The scales or indices were constructed from the factor analysis of the items contained in the questionnaires. The indices were based on theoretical concepts, on the findings of my previous research (Bradley, 1972), and the study of the Postal Giro Service

(Chapter 4). The response alternatives to the questions were graded from one to five. Some of the response alternatives in the section 'Information and co-determination' in the questionnaire used were graded from one to three. However, in contrast to the 'yes/no' questions that were eliminated from the factor analyses, these were on a graduated scale. In some cases the response scale has been inverted so that five (or three in the 1–3 scale) always indicated a positive attitude and one indicated a negative attitude. Because of the nature of the questions, this approach could not be applied to the indices related to patterns of values and a general attitude to work (the V indices) i.e. 'weighting, internal', 'weighting, external' and 'weighting, development'.

An average index value was obtained by totalling the points for each question in the index for each individual, and then dividing the total by the number of questions. If a respondent had not answered every question in the index, then that index was only included if at least half of the questions plus one had been answered. The average index value could vary between 1·0 and 5·0 (and 1·0 and 3·0 for the exception mentioned above). The reliability of all indices had been estimated using Cronbach's alpha formula (Cronbach, 1951).

Our comparisons included terminal users and non-terminal users with regard to different 'blocks' or problem areas:

- the psychosocial aspects of the work environment (18 indices);
- perceived effects of the terminal system and computerization in general (14 indices);
- aspects of democracy at the workplace (5 indices);
- health aspects (11 indices).

We also compared groups in the same 'blocks' that had different lengths of experience of working on VDTs, groups with different daily working hours at the VDT, and groups with varying amounts of customer contact with or without 'computer contact' at the same time. Only statistically reliable differences are presented below.

A summary of the quantitative measurements and interviews before computerization

Some experience was gained from phases in earlier work, namely from the measurements before the introduction of the data processing system and interviews in 1976 (points 1 and 3 above). The research project at Trygg-Hansa began with a 'quantitative measurement' before the data processing system was introduced (point 1 above) aimed at all personnel at the head office. Its aim was to study employees' expectations and fears about the new terminal system. Some departments feared that the introduction of the new terminal system would have some negative effects on their work environment. These departments were the Accounts Department, Private Property Insurance, Personal Insurance, Personal Insurance Payments and Company and Industrial Insurance.

What were the differences between these departments and other departments in the organization? With the exception of the Personal Insurance Department, these were the departments in which terminals were to be installed in stages from 1977 onwards. The installation plan also included installing terminals in the Customer Service Department and the Claims Department in Stockholm. However, both these departments had been using query-type terminals for some time. This experience of using VDTs may possibly have helped lessen the anxiety felt prior to the introduction of VDTs. The difference between the work of the Customer Service Department and the Claims Offices and other departments is that these offices have both telephone and personal contacts with customers.

The interviews with users of dialogue terminals (point 3 above) gave the following results. The people interviewed had used the terminal system for about four months. Most of the users who had experience of working on VDTs reacted positively to the change. Some of them reported a marked improvement in their work situation. For example, the old system of writing everything down had been considered laborious. A couple of the staff were suffering industrial injuries brought on by twenty years of using one hand to record data. The terminals allowed them to use both hands. The keys were also easier to press, and the physical load was evenly distributed over the body. Other improvements were that they thought they had come closer to the 'source' of the work, in other words, they were no longer just one stage in a process; and they felt that the VDT made work easier in purely physical terms (head, neck and hands); the text on the screen was sharp and easy to read compared to the printed form; they could get a better total picture of the work; and the work itself was more enjoyable and generally easier to perform.

Opinions were divided on the question of the risk of error in the work. On the one hand, not having a pile of papers to refer back to during the day created a feeling of uncertainty; on the other hand, it was now possible to locate errors on the VDT instead of using the error lists and this made it easier for the individual worker to correct mistakes. Some of the people interviewed felt that computer breakdowns and waiting times meant that they could no longer set their own pace of work. However, this occurred mostly in the early stages.

A crucial factor with all technical aids is the extent that the operator is confined to the work station and the physio-ergonomic problems and social isolation this causes. We asked ourselves if the operator was not only 'forced' to sit at the machine, but whether the automatic equipment also had a psychologically limiting effect on personnel? The interviews indicated that there were some trends in this direction: staff felt anxious about leaving their 'equipment'; because work had to be planned, spontaneous contact with co-workers was rather more difficult (the VDT had to be turned off); and staff telephoned each other instead of walking over to talk; VDTs require telephones, and this increased the amount of 'remote contact'. The people interviewed did not see the computers as a direct threat. There had been speculation about staff cuts in all previous reorganizations—either mergers, or the introduction of the open-plan office layout—but here it was felt that different work tasks had emerged instead. However, several of the people interviewed wondered whether computerization would not lead to a loss of jobs and 'marginalization' at workplaces—as they said, 'not here, but at other workplaces'.

There was some conflict at Trygg-Hansa between two forms of rationalization: between computerization and the introduction of an open-plan office layout. Screens were fitted with steel hoods to shield them from direct light. These hoods were considered to give good protection from daylight. But operators were also irritated by the glare on the keyboards from the overhead lighting. They said that the open-plan office had lost its feeling of space and airiness since the screens had been installed.

Some of the people interviewed felt that installing the VDTs had reduced the opportunity for contact and communication. This may be partly because the work requires more attention. It is probably also a question of the psychological effects of the mechanical equipment. The hoods around the VDTs blocked forward vision, adding to the operators' isolation. The terminals also meant that the operators handled more cases, often increasing the pace of work, and they no longer needed to leave their work stations to fetch information.

Some of the people interviewed reported fears of long-term hazards in working on VDTs such as:

- damage to the sight;
- the risk of radiation from the screen;
- neck and back complaints and other physical complaints caused by an unsuitable working position;
- cramped working conditions: the VDT takes up a lot of room;
- fewer opportunities to make individual modifications to the work routines;
- the work becomes mechanical;

- human resources are utilized less and less;
- users are more confined to the work station;
- more attention is required.

The results from these introductory interviews should be viewed in the light of the fact that the interviewees had used VDTs for a mere four months. The staff were enthusiastic about a number of points, but they were anxious and doubtful about several other issues. People's 'adaptation mechanisms' had hardly had time to come into effect. At the same time it was impossible at this phase of the research project to assess the more permanent or long-term psychosocial effects of introducing VDTs. The results of this phase of the project are presented in the report, 'Arbetsmiljö och terminaler: Inför och under en förändring' ('Work Environment and Display Terminals: Before and During a Process of Change'), RAM internal report, Bradley, 1979. This report emphasized the importance of heeding the demands on the working environment that stemmed from the need for cooperation and fellowship among people, that is, problems related to human communication. In the future, more and more people will communicate with the computer instead of with other people.

The use of VDTs brings about qualitative and quantitative changes in personal contacts, for example, because of the requirement for attentiveness and the way terminals confine the operator to his/her workplace. This creates both physio-ergonomic problems and social isolation. A telephone—which must be 'watched'—is a common complement to the VDT. This creates a risk of depleting the content of human communication, and communication helps give work its value (Bradley, 1979).

The quantitative follow-up survey made it possible to arrive at more reliable conclusions about introducing VDTs to the workplace. These measurements were taken in late 1978 and early 1979 (point 6 above). We focused on the following factors:

- the psychosocial aspects of the work environment;
- perceived effects of the terminal system and of computerization in general;
- aspects of democracy at the workplace;
- health factors;
- stress factors in the work.

We compared:

- terminal users and non-terminal users;
- groups with differing lengths of experience of terminal use;
- groups that had differing lengths of daily terminal use;

- groups that used different types of terminal system;
- personnel with varying degrees of contact with the general public;
- secretaries and other female personnel;
- different departments.

Work Environment and Display Terminals (Bradley, 1981) describes the various methodical processes that preceded the results reported in the next section. This book also includes the complete statistical tables of material on which the findings were based.

3. Results

Figure 4 is a general diagram of the variables that were used in the survey. The indices are included, as are the background variables that were included in the survey. The abbreviations in the groups of indices are also to be found in Tables 5.2–5.8. The indices and background data have been arranged so that they relate to the original theoretical model of the project (see Chapter 3, Figures 1 and 2). The following groups of variables are to be found in Figure 4:

- work-related background data
- individual-related background data
  - personal data
  - family circumstances
- weighting of work environment factors and general attitude to work (V indices)
- work environment factors (perceived work environment conditions)
  - psychosocial work environment factors (S indices)
  - democracy at work (DEM indices)
  - physio-ergonomic work environment factors (F indices)
- the effects of the data processing system on the work environment
  - effects of the terminal system (T indices)
  - general effects of the data processing system (D indices)
- the effects of a bad/good work environment
  - health problems at work (HA indices)
  - health problems in leisure time (HF indices)
  - the effects of work on leisure time/family life (AF indices)
  - the effects of the data processing system on leisure time activities (DF indices).

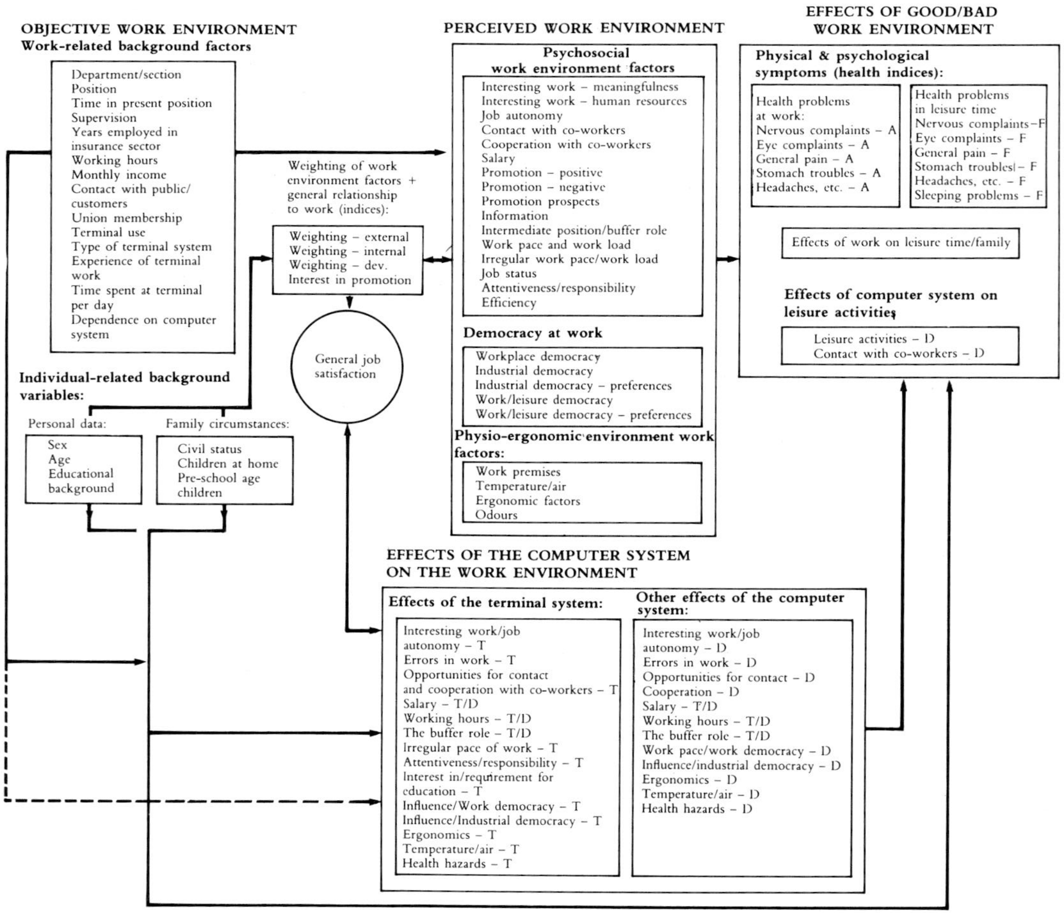

Figure 4. A chart of the indices (from Bradley's model)

Source: Bradley (1981)

A list of all indices related to the models used in the investigation is to be found in Appendix 1. The abbreviations for indices and the indices described in composite questions are also included.

Different 'blocks' of problems were analyzed in relation to the project's theoretical model:

- the psychosocial aspects of the work environment;
- the effects of the VDT system and the general impact of computerization;
- aspects of democracy at the workplace;
- health factors.

Different comparisons were made when analyzing these problem areas. Analysis of the mean differences was made with the help of T-tests. This material can be analyzed in many ways. On this occasion, a number of processes were selected which it was considered should be given priority with regard to the prevailing level of knowledge. Specific hypotheses may subsequently be advanced and tested.

Tables 5.2–5.12 compare the mean values in the total group; values were then determined separately for men, women and younger and older personnel. It was shown that the use of terminals did, to some degree, correlate with the sex and age variables. The mean values for the groups between which there were significant differences in mean values at the 5 per cent level ($p<0.05$) are reported. Significant differences at the 1 per cent level are indicated with a minus sign (−) between the mean values. Other differences in values where $p = 0.05$ have been included and the mean values have been stated in brackets. These mean values have been included only to illustrate some trends. Some of the mean value comparisons occur several times in different tables, namely those between men and women and between younger and older personnel. This is only intended to be of help in interpreting significant mean value differences in the subsequent reports on the findings.

The following age classification was used:

Younger: All respondents under the age of 48;

Older: All respondents over the age of 47;

Total: Placed above the columns which include all people in the investigation.

Tables 5.2–5.4 report the differences between non-terminal users (non-T) and terminal users (Term). The group of terminal users includes data entry personnel, which is a very small group in this investigation. The mean values for the entire statistical sample, broken down by sex and age, are reported in the columns Women/Men and Younger/Older.

Table 5.2 Significant mean differences (t-test): a comparison of non-terminal users with terminal users

ABBR. INDEX	INDEX	TOTAL non-T Term	WOMEN non-T Term	WOMEN/ MEN	MEN non-T Term	YOUNGER non-T Term	YOUNGER/ OLDER	OLDER non-T Term
SIME	Interesting work – meaningful		4.03 4.32	4.17 – 4.39				
SIMR	Interesting work – human resources		(3.18) (3.49)	3.35 – 3.80	3.97 3.45			
SSJA	Job autonomy		3.36 3.64	3.50 – 4.09	4.24 – 3.70			
SKONT	Contact opportunities at work			3.46 – 4.12				
SSAM	Cooperation with co-workers							
SLÖN	Salaries				3.03 2.57		2.80 3.06	
SBEFP	Promotion – positive	4.27 4.11	4.32 4.11				4.10 – 4.29	
SBEFN	Promotion – negative							
SBEFM	Promotion prospects			3.03 3.24				
SINFO	Information							
SBUFF	Intermediate position/buffer role			3.51 3.77			3.48 3.74	
SARBT	Work pace/work load	3.09 – 2.78			3.14 2.76			3.12 2.74
SOJA	Irregular work pace/work load	2.94 – 2.63						3.04 – 2.60
SAANS	Job status			3.45 – 3.74			3.44 3.69	
SUPPA	Attentiveness/responsibility			2.88 3.18				
SEFF	Efficiency							
DEMA	Work democracy			2.58 – 2.92	3.04 – 2.64			2.82 2.58
DEMF	Company democracy	1.29 1.19		1.15 – 1.40	1.44 1.28	1.25 1.13	1.20 1.30	
DEMFÖ	Company democracy – wishes						1.60 1.72	
DEMAL	Work/leisure time democracy	2.59 2.44		2.34 – 2.77				
DEMALÖ	Work/leisure time democracy – wishes	1.96 – 1.86		1.88 – 1.97	2.00 1.92			1.96 1.87
FA	Work premises							
FT	Temperature/air			2.24 – 3.17				2.77 2.28
FE	Ergonomic factors							
FL	Odours			3.37 – 3.92				3.77 3.23

Level of significance:
$p < 0.05$
minus (–): $p < 0.01$
brackets (): $p = 0.05$

Table 5.3 Significant mean differences (t-test): a comparison of non-terminal users with terminal users

ABBR. INDEX	INDEX	TOTAL non-T Term	WOMEN non-T Term	WOMEN/ MEN	MEN non-T Term	YOUNGER non-T Term	YOUNGER/ OLDER	OLDER non-T Term
TISJA	Interesting work/job autonomy – T							
TFEL	Errors in work – T							
TKONTSAM	Contact opportunities with co-workers – T					2.99 – 3.07		
TDBUFF	Intermediate position/buffer role – T/D	2.95 – 2.73	2.97 – 2.74		2.93 – 2.72	2.92 –2.76		2.97 – 2.69
TINFLAD	Influence – work democracy – T							
TINFLFD	Influence – company democracy – T							
TOJA	Irregular work pace – T	2.97 – 2.61	2.97 – 2.67		2.98 – 2.48	3.02 – 2.67		2.94 – 2.53
TUPPA	Attentiveness – responsibility – T	2.95 – 2.64	2.98 – 2.67		2.93 – 2.58	3.03 – 2.62		2.89 2.67
TDLÖN	Wages – T/D							
TDARBTID	Working hours – T/D							
TUTIK	Interest in education – requirements – T	3.37 – 2.98	3.40 – 2.97					3.41 3.00
TERGO	Ergonomic factors – T							
TTEMPL	Temperature/air – T	3.01 2.85	3.03 – 2.78	2.88 3.01		3.03 2.87		
THÄLSOR	Health hazards – T					2.62 2.25		
DISJA	Interesting work and job autonomy – D			2.91 – 3.18				
DEFEL	Errors in work – D							
DKONT	Contact opportunities – D							
DSAM	Cooperation – D		2.99 3.08					
DINFLAD	Influence – work democracy – D			2.83 2.98				
DINFLFD	Influence – company democracy – D						3.00 2.91	
DATAB	Work pace/work load – D	2.80 – 2.50	2.70 2.42	2.59 – 2.80		2.82 2.61		2.78 – 2.35
DUTIK	Interest in education – requirements – D			3.25 – 2.96				
DERGO	Ergonomic factors – D							
DTEMPL	Temperature/air – D			2.90 3.02				
DHÄLSOR	Health hazards – D							

Level of significance:
$p < 0.05$
minus (–): $p < 0.01$
brackets (): $p = 0.05$

Table 5.4 Significant mean differences (t-test): a comparison of non-terminal users with terminal users

ABBR. INDEX	INDEX	TOTAL	WOMEN	WOMEN/ MEN	MEN non-T Term	YOUNGER non-T Term	YOUNGER/ OLDER	OLDER non-T Term
HAN	Nervous complaints – A			3.52 – 3.87				
HAÖ	Eye complaints – A			3.43 – 3.98		3.88 3.50		
HAV	General pain – A	3.97 – 3.61		3.54 – 4.29		4.32 – 3.56		
HAM	Stomach complaints – S							
HAH	Headaches, etc. – A	4.00 – 3.63		3.69 – 4.13		4.00 3.60		
HFN	Nervous complaints – F			3.51 – 3.84		3.72 3.46		
HFÖ	Eye complaints – F			3.54 – 4.04				
HFV	General pain – F	4.13 3.88		3.81 – 4.38		4.42 – 3.87	4.17 3.94	
HFM	Stomach complaints – F							
HFH	Headaches, etc. – F	4.12 3.80		3.81 – 4.26				
HFS	Sleeping problems – F			4.16 4.40			4.41 – 4.13	
AF	Effects of work on leisure time/family life			4.21 3.91	3.72 4.25		3.96 4.20	4.09 4.39
DFSYM	Leisure time activities – D							
DFKM	Contact – D			2.99 3.03				
VY	Weighting – external			2.30 – 2.76				
VI	Weighting – internal			2.28 – 2.03				
VU	Weighting – development			2.96 – 2.63			2.63 – 3.00	
VB	Interest in promotion	2.80 3.10			2.76 – 3.42		3.37 – 2.60	2.41 2.75

Level of significance:
$p < 0.05$
minus (–): $p < 0.01$
brackets (): $p = 0.05$

Table 5.5 Significant mean differences (t-test): a comparison of terminal users spending varying lengths of time daily at the VDU

ABBR. INDEX	INDEX	TERMINAL USERS <30mins	TERMINAL USERS >2hrs	WOMEN <30mins	WOMEN >2hrs	WOMEN/ MEN	MEN <30mins	MEN >2hrs	YOUNGER <30mins	YOUNGER >2hrs	YOUNGER/ OLDER	OLDER <30mins	OLDER >2hrs
SIME	Interesting work – meaningful					4.17 – 4.39							
SIMR	Interesting work – human resources					3.35 – 3.80	(4.00)	(3.33)				(3.79)	(3.11)
SSJA	Job autonomy					3.50 – 4.09							
SKONT	Contact opportunities at work					3.46 – 4.12							
SSAM	Cooperation with co-workers												
SLÖN	Salaries										2.80 3.06		
SBEFP	Promotion – positive	4.23	3.92	4.27	3.96				4.27	3.88	4.10 – 4.29		
SBEFN	Promotion – negative											2.71	2.17
SBEFM	Promotion prospects					3.03 3.24							
SINFO	Information	3.66 –	2.93	3.62	2.96		3.77	2.87	3.61	2.93		(3.73)	(2.93)
SBUFF	Intermediate position/buffer role					3.51 3.77					3.48 3.74		
SARBT	Work pace/work load											2.87 –	2.03
SOJA	Irregular work pace/work load												
SAANS	Job status					3.45 – 3.74					3.44 3.69		
SUPPA	Attentiveness/responsibility					2.88 3.18							
SEFF	Efficiency												
DEMA	Work democracy					2.58 – 2.92							
DEMF	Company democracy					1.15 – 1.40					1.20 1.30		
DEMFÖ	Company democracy – wishes										1.60 1.72		
DEMAL	Work/leisure time democracy					2.34 – 2.77							
DEMALÖ	Work/leisure time democracy – wishes					1.88 – 1.97							
FA	Work premises												
FT	Temperature/air					2.24 – 3.17							
FE	Ergonomic factors												
FL	Odours	3.38	4.05			3.37 – 3.92							

Level of significance:
$p < 0.05$
minus (–): $p < 0.01$
brackets (): $p = 0.05$

Table 5.6 Significant mean differences (t-test): a comparison of terminal users spending varying lengths of time daily at the VDU

ABBR. INDEX	INDEX	TERMINAL USERS		WOMEN		WOMEN/ MEN		MEN		YOUNGER		YOUNGER/ OLDER		OLDER	
		<30mins	>2hrs	<30mins	>2hrs			<30mins	>2hrs	<30mins	>2hrs			<30mins	>2hrs
TISJA	Interesting work/job autonomy – T														
TFEL	Errors in work – T	2.87 –	3.40	2.76	3.36									2.75	4.00
TKONTSAM	Contact opportunities with co-workers – T													3.05	3.31
TDBUFF	Intermediate postion/buffer role – T/D	2.82	2.48	2.87	2.46					2.83	2.49				
TINFLAD	Influence – work democracy – T														
TINFLFD	Influence – company democracy – T														
TOJA	Irregular work pace – T														
TUPPA	Attentiveness – responsibility – T														
TDLÖN	Wages – T/D														
TDARBTID	Working hours – T/D	2.69	3.13	2.58	3.05									2.80	3.65
TUTIK	Interest in education – requirements – T	3.09	2.63												
TERGO	Ergonomic factors – T	3.03	3.25	2.97	3.20									(3.05)	(3.50)
TTEMPL	Temperature/air – T					2.88	3.01								
THÄLSOR	Health hazards – T														
DISJA	Interesting work and job autonomy – D					2.91 –	3.18								
DEFEL	Errors in work – D	2.70	3.25												
DKONT	Contact opportunities – D														
DSAM	Cooperation – D														
DINFLAD	Influence – work democracy – D					2.83	2.98							2.67	3.07
DINFLFD	Influence – company democracy – D			(2.86)	(3.07)							3.00	2.91		
DATAB	Work pace/work load – D			2.54	2.03	2.59 –	2.80	2.38	3.19						
DUTIK	Interest in education – requirements – D					3.25 –	2.96								
DERGO	Ergonomic factors – D														
DTEMPL	Temperature/air – D					2.90	3.02								
DHÄLSOR	Health hazards – D														

Level of significance:
$p < 0.05$
minus (–): $p < 0.01$
brackets (): $p = 0.05$

Table 5.7 Significant mean differences (t-test): a comparison of terminal users spending varying lengths of time daily at the VDU

ABBR. INDEX	INDEX	TERMINAL USERS	WOMEN etc.	WOMEN/ MEN	MEN	YOUNGER	YOUNGER/ OLDER	OLDER
		<30mins >2hrs	<30mins >2hrs		<30mins >2hrs	<30mins >2hrs		<30mins >2hrs
HAN	Nervous complaints – A			3.52 – 3.87				
HAÖ	Eye complaints – A			3.43 – 3.98				
HAV	General pain – A	3.73 3.14		3.54 – 4.29	4.35 – 3.15			
HAM	Stomach complaints – S							
HAH	Headaches, etc. – A			3.69 – 4.13				
HFN	Nervous complaints – F			3.51 – 3.84				
HFÖ	Eye complaints – F			3.54 – 4.04				
HFV	General pain – F			3.81 – 4.38			4.17 3.94	
HFM	Stomach complaints – F							
HFH	Headaches, etc. – F			3.81 – 4.26		3.64 – 4.46		
HFS	Sleeping problems – F			4.16 4.40			4.41 – 4.13	
AF	Effects of work on leisure time/family life			4.21 3.91			3.96 4.20	
DFSYM	Leisure time activities – D				3.02 3.15			3.00 3.10
DFKM	Contact – D			2.99 3.03				
VY	Weighting – external			2.30 – 2.76				
VI	Weighting – internal			2.28 – 2.03				
VU	Weighting – development			2.96 – 2.63			2.63 – 3.00	
VB	Interest in promotion						3.37 – 2.60	

Level of significance:
$p < 0.05$
minus (–): $p < 0.01$
brackets (): $p = 0.05$

Table 5.8 Significant mean differences (t-test): a comparison of groups with different lengths of experience of terminal use

ABBR. INDEX	INDEX	TERMINAL USERS 1m–1y	TERMINAL USERS >1y	WOMEN 1m–1y	WOMEN >1y	WOMEN/ MEN	MEN 1m–1y	MEN >1y	YOUNGER 1m–1y	YOUNGER >1y	YOUNGER/ OLDER	OLDER 1m–1y	OLDER >1y
SIME	Interesting work – meaningful					4.17 – 4.39							
SIMR	Interesting work – human resources			3.08	3.65	3.35 – 3.80							
SSJA	Job autonomy					3.50 – 4.09							
SKONT	Contact opportunities at work					3.46 – 4.12							
SSAM	Cooperation with co-workers												
SLÖN	Salaries										2.80 3.06		
SBEFP	Promotion – positive										4.10 – 4.29		
SBEFN	Promotion – negative												
SBEFM	Promotion prospects	2.81	3.23			3.03 3.24							
SINFO	Information												
SBUFF	Intermediate position/buffer role	3.90	3.39			3.51 3.77			3.90 –	3.26	3.48 3.74		
SARBT	Work place/work load	3.30 –	2.62	3.42 –	2.60				3.40 –	2.50			
SOJA	Irregular work pace/work load	3.06	2.59				3.33	2.62					
SAANS	Job status			3.08	3.64	3.45 – 3.74					3.44 3.69		
SUPPA	Attentiveness/responsibility	3.46 –	2.82	3.38	2.69	2.88 3.18			3.57 –	2.72			
SEFF	Efficiency												
DEMA	Work democracy					2.58 – 2.92							
DEMF	Company democracy					1.15 – 1.40					1.20 1.30		
DEMFÖ	Company democracy – wishes										1.60 1.72		
DEMAL	Work/leisure time democracy	2.65 –	2.33			2.34 – 2.77	2.87	2.60	2.76 –	2.28			
DEMALÖ	Work/leisure time democracy – wishes					1.88 – 1.97			1.98 –	1.80			
FA	Work premises			2.33	3.13								
FT	Temperature/air					2.24 – 3.17							
FE	Ergonomic factors			3.90	4.22								
FL	Odours					3.37 – 3.92							

Level of significance:
p < 0.05
minus (–): p < 0.01
brackets (): p = 0.05

Table 5.9 Significant mean differences (t-test): a comparison groups with different lengths of experience of terminal use

ABBR. INDEX	INDEX	TERMINAL USERS 1m–1y	TERMINAL USERS >1y	WOMEN 1m–1y	WOMEN >1y	WOMEN/MEN		MEN 1m–1y	MEN >1y	YOUNGER 1m–1y	YOUNGER >1y	YOUNGER/OLDER		OLDER 1m–1y	OLDER >1y
TISJA	Interesting work/job autonomy – T														
TFEL	Errors in work – T														
TKONTSAM	Contact opportunities with co-workers – T														
TDBUFF	Intermediate position/buffer role – T/D														
TINFLAD	Influence – work democracy – T														
TINFLFD	Influence – company democracy – T														
TOJA	Irregular work pace – T														
TUPPA	Attentiveness – responsibility – T														
TDLÖN	Wages – T/D							3.21	2.92	(3.21)	(3.01)				
TDARBTID	Working hours – T/D														
TUTIK	Interest in education – requirements – T														
TERGO	Ergonomic factors – T														
TTEMPL	Temperature/air – T					2.88	3.01	(2.85)	(3.19)						
THÄLSOR	Health hazards – T														
DISJA	Interesting work and job autonomy – D					2.91 –	3.18							2.76 –	3.22
DEFEL	Errors in work – D														
DKONT	Contact opportunities – D													2.96	3.14
DSAM	Cooperation – D													2.96	3.20
DINFLAD	Influence – work democracy – D					2.83	2.98								
DINFLFD	Influence – company democracy – D											3.00	2.91		
DATAB	Work pace/work load – D					2.59 –	2.80								
DUTIK	Interest in education – requirements – D					3.25 –	2.96								
DERGO	Ergonomic factors – D	(3.00)	(3.13)												
DTEMPL	Temperature/air – D					2.90	3.02								
DHÄLSOR	Health hazards – D														

Level of significance:
$p < 0.05$
minus (–): $p < 0.01$
brackets (): $p = 0.05$

Table 5.11 *Significant mean differences (t-test): a comparison between groups using different types of terminal system, and with different amounts of contact with the public*

		TYPE OF TERMINAL SYSTEM			CONTACT WITH THE PUBLIC					
		TERMINAL USER			TOTAL			TERMINAL USER		
ABBR. INDEX	INDEX	Query system		Dialogue system	Often Sometimes		Seldom Never	Often Sometimes		Seldom Never
SIME	Interesting work – meaningful				4.40	–	4.13			
SIMR	Interesting work – human resources				3.73	–	3.35			
SSJA	Job autonomy				3.91	–	3.59			
SKONT	Contact opportunities at work									
SSAM	Cooperation with co-workers									
SLÖN	Salaries				2.83		3.04			
SBEFP	Promotion – positive									
SBEFN	Promotion – negative									
SBEFM	Promotion prospects									
SINFO	Information									
SBUFF	Intermediate position/buffer role				3.47	–	3.78	3.35	–	3.89
SARBT	Work pace/work load				2.75	–	3.21	2.59	–	3.07
SOJA	Irregular work pace/work load							2.45	–	2.89
SAANS	Job status									
SUPPA	Attentiveness/responsibility							2.75		3.24
SEFF	Efficiency									
DEMA	Work democracy				2.80		2.65			
DEMF	Company democracy									
DEMFÖ	Company democracy – wishes									
DEMAL	Work/leisure time democracy	2.31	–	2.64						
DEMALÖ	Work/leisure time democracy – wishes									
FA	Work premises									
FT	Temperature/air							2.16	–	2.85
FE	Ergonomic factors	4.22		3.92						
FL	Odours									

TISJA	Interesting work/job autonomy – T									
TFEL	Errors in work – T									
TKONTSAM	Contact opportunities with co-workers – T									
TDBUFF	Intermediate position/buffer role – T/D				2.79	–	2.91	2.67		2.83
TINFLAD	Influence – work democracy – T									
TINFLFD	Influence – company democracy – T									
TOJA	Irregular work place – T									
TUPPA	Attentiveness – responsibility – T									
TDLÖN	Wages – T/D				2.98	–	3.14	2.96	–	3.18
TDARBTID	Working hours – T/D									
TUTIK	Interest in education – requirements – T									
TERGO	Ergonomic factors – T									
TTEMPL	Temperature/air – T									
THÄLSOR	Health hazards – T									
DISJA	Interesting work and job autonomy – D									
DEFEL	Errors in work – D									
DKONT	Contact opportunities – D				2.99		3.06			
DSAM	Cooperation – D									
DINFLAD	Influence – work democracy – D							2.76		3.07
DINFLFD	Influence – Company democracy – D							2.88		3.01
DATAB	Work pace/work load – D									
DUTIK	Interest in education – requirements – D									
DERGO	Ergonomic factors – D									
DTEMPL	Temperature/air – D	2.83		3.07						
DHÄLSOR	Health hazards – D				2.43		2.64			

Level of significance:
$p < 0.05$
minus sign (–): $p < 0.01$
brackets (): $p = 0.05$

Table 5.10 Significant mean differences (t-test): a comparison of groups with different lengths of experience of terminal use

ABBR. INDEX	INDEX	TERMINAL USERS 1m–1y >1y	WOMEN 1m–1y >1y	WOMEN/ MEN	MEN 1m–1y >1y	YOUNGER 1m–1y >1y	YOUNGER/ OLDER	OLDER 1m–1y >1y
HAN	Nervous complaints – A			3.52 – 3.87		3.86 3.41		
HAÖ	Eye complaints – A			3.43 – 3.98				
HAV	General pain – A			3.54 – 4.29				
HAM	Stomach complaints – S							
HAH	Headaches, etc. – A			3.69 – 4.13				
HFN	Nervous complaints – F	(3.79) (3.48)		3.51 – 3.84		3.82 – 3.27		
HFÖ	Eye complaints – F			3.54 – 4.04				
HFV	General pain – F	4.22 3.80		3.81 – 4.38		4.37 – 3.67	4.17 3.94	
HFM	Stomach complaints – F							
HFH	Headaches, etc. – F			3.81 – 4.26				
HFS	Sleeping problems – F			4.16 4.40		4.71 4.21	4.41 – 4.13	
AF	Effects of work on leisure time/family life			4.21 3.91			3.96 4.20	
DFSYM	Leisure time activities – D		3.00 2.97					
DFKM	Contact – D			2.99 3.03				
VY	Weighting – external			2.30 – 2.76				
VI	Weighting – internal			2.28 – 2.03				
VU	Weighting – development			2.96 – 2.63			2.63 – 3.00	
VB	Interest in promotion	3.45 3.01	3.46 2.88				3.37 – 2.60	3.50 2.65

Level of significance:
$p < 0.05$
minus (–): $p < 0.01$
brackets (): $p = 0.05$

Table 5.12 Significant mean differences (t-test): a comparison between groups using various types of terminal system, and with different amounts of contact with the public

ABBR. INDEX	INDEX	TYPE OF TERMINAL SYSTEM TERMINAL USER		CONTACT WITH THE PUBLIC TOTAL		CONTACT WITH THE PUBLIC TERMINAL USER	
		Query system	Dialogue system	Often Sometimes	Seldom Never	Often Sometimes	Seldom Never
HAN	Nervous complaints – A						
HAÖ	Eye complaints – A						
HAV	General pain – A						
HAM	Stomach complaints – S						
HAH	Headaches, etc. – A						
HFN	Nervous complaints – F						
HFÖ	Eye complaints – F						
HFV	General pain – F						
HFM	Stomach complaints – F						
HFH	Headaches, etc. – F						
HFS	Sleeping problems – F						
AF	Effects of work on leisure time/family life						
DFSYM	Leisure time activities – D						
DFKM	Contact – D						
VY	Weighting – external						
VI	Weighting – internal			2.06 –	2.28		
VU	Weighting – development			2.71	2.92		
VB	Interest in promotion						

Level of significance:
$p < 0.05$
minus sign (−): $p < 0.01$
brackets (): $p = 0.05$

The differences between people with different lengths of daily terminal use are set out in Tables 5.5–5.7. The designation <30m/d signifies terminal users who use a VDT for less than thirty minutes a day and >2h/d signifies terminal users who use a VDT for more than two hours a day.

The differences between terminal users' length of experience of terminal work are reported in Tables 5.8–5.10. The Women/Men and Younger/Older columns represent the general differences between the sexes and age groups without considering the degree of contact with the terminal system.

The differences between terminal users who work on different types of terminal system, both query systems and dialogue systems, are reported in Tables 5.11–5.12 in the column headed 'Type of Terminal System'. We use 'dialogue system' in the accepted use of the term, and not the company's term for its data processing system. Here, we mean a system for entering and retrieving information.

Under the heading 'Contact with the General Public', the differences between people in the sample who report contact with the general public often or sometimes, and those who report contact with the general public seldom or never are recorded under the heading 'Total'. Comparisons were also made between corresponding groups among terminal users (under 'Terminal Users').

Working on display terminals: a comparison between non-terminal users and terminal users

The report of the findings follows the concepts stated in the theoretical models in Figures 1 and 2 in Chapter 3, and in the 'empirical' model set out in Figure 4, where the investigation variables (the index level) are grouped under the headings in the boxed fields in the models. The text describes the significant differences found between the groups that were compared. In these comparisons, age and sex were kept as constants.

The psychosocial work environment

The differences in the total sample are of particular interest in terms of the areas in the psychosocial work environment (see the list of indices in the model in Figure 4). Factors concerning work pace, staffing, efficiency requirements, etc. are the most prominent areas in which terminal users expressed a higher level of dissatisfaction than non-users. But the pattern does not hold when the sex factor is taken into account. For both sexes, the 'terminal-sensitive' areas of the work revolve around its content, with particular emphasis on the degree of

Some departments, e.g., Customer Services, use display units and the telephone for simultaneous communication.

Photo: Gunnar Wåhlén

job autonomy the user has and how interesting the job content is. Among women, however, terminal users are more positive than non-users to their work content. The reverse is true for men.

One explanation of these findings may be that according to the traditional allocation of roles at workplaces, the work tasks of male white-collar workers are often more interesting and offer a greater degree of job autonomy than those of female employees in the same category, and male white-collar workers are more used to receiving completed documentation. Women find that producing their own material and basic documentation is a natural part of their work and they have now been given an 'aid' in this work. To some extent, this interpretation is supported by the results of the comparisons between men and women in the total group. There is no difference in the total sample between men's and women's attitudes to *the effects of the terminal system*. However, the *computer system in general* is seen more often by female than by male personnel as contributing to a worsened work situation. It is likely that the changes in the structure and organization of work in the company also had some effect (see Chapter 9). Comparisons between the sexes on all indices are examined in the penultimate section of this chapter. At many workplaces, computerization has helped to reinforce the principles that dominate the allocation of work, thus constituting a threat to the attempt to achieve equality between men and women at work. This is true of both the so-called external and internal aspects of equality (see Chapter 9).

A comparison may be made here with the findings of the Postal Giro Service investigation (see Chapter 4). In this investigation, it was mostly the personnel in the organization's operating departments (mainly bookkeeping) and the Data Processing Department who reported various work environment problems. Most of the Data Processing Department's staff were young, highly qualified men, many of them working on shift, and who had been employed for a relatively short time. All the personnel in the operations or bookkeeping departments were women with relatively poor qualifications, a high average age, and who had been working there for a long time. Half of them worked part-time.

Democracy at the workplace

The terminal users expressed more dissatisfaction about democracy at the workplace (see the box in the model, Figure 4). They mentioned having influence on issues of more general interest to the company, having influence on scheduling breaks, holidays and working hours, and finally a wish to gain greater influence in the working hours issues mentioned above. The older terminal users emphasized temperature and ventilation problems.

The effects of the terminal system

The terminal users pointed out ways in which computerization had worsened the work environment (see 'Effects of the Terminal System' in the model). The buffer role (the demands placed upon personnel at work), the irregular pace of work, the need for attentiveness and the amount of responsibility at work were the work environment factors that had worsened or had become harder to deal with. The increase in buffer role problems may be attributable to the more rapid throughput of work that resulted from introducing new technology, which has also increased both the demands people place upon themselves and demands that originate in their surroundings, for instance from customers. Customers assume that personnel have access to a computer terminal, so they expect rapid response times. They used to be given this type of information by telephone on the following day.

Further comments on the work environment phenomenon known as 'the intermediate position—the buffer role' are relevant here. These comments will include quotations from *'Yrkesroller—livsmiljö' (Working Roles and the Quality of Life,* Bradley, 1979). The action taken on the buffer role is discussed in more detail in Chapter 9. The problem of the psychosocial aspects of the intermediate position had begun to emerge as a serious work environment problem in, for example, the social services and education sectors—professions that involve a coordinating function in an increasingly complicated work situation. But there are also inherent problems in the way that work is normally organized.

The intermediate position can be found in many situations:

- between superior and subordinate within an organization;
- between different departments/groups within an organization;
- in relation to other departments;
- in relation to other companies/organizations;
- in relation to the general public/customers/clients;
- in relation to private life;
- in relation to one's own 'internal' demands.

When people interact, most 'demands' work both ways. It is often the individual that constitutes the 'environment' for one or more other individuals. He/she thus helps to create a potential intermediate position and an indirect buffer role for other individuals and groups either within or outside his/her organization. Sometimes these different types of intermediate position are related to one and the same position. In some work situations, one form of intermediate position may become more pronounced. Conflicting demands from different sources will generate conflicting roles, from which we derive the term 'the buffer role'.

Because being involved with—and living in—different roles will enrich the individual, it is in itself an asset. But this situation can also be far too great a load for the individual to bear if the demands and expectations of the different roles are contradictory, and the individual may develop various defensive attitudes which may affect his/her psychological health (see Chapter 9).

Finally, terminal users say that the introduction of the terminal system had some influence on personal interest in education and on the company's requirement of training/further training. The younger terminal users often claim that the terminals have increased the risk of physical and psychological problems.

Health

The health issues—'physical and psychological symptoms'—are at the far right of the model in Figure 4. Fewer non-terminal users than terminal users reported complaints such as general pain or headaches during working hours and pain or headaches during their leisure time. In the younger group, terminal users report that they are sometimes troubled by eye complaints; non-terminal users have a lower incidence of eye complaints. By contrast, however, there are no differences within the total group between terminal and non-terminal users. Efforts have been made to avoid work involving long periods of continuous use of the VDTs. Initially, there was a certain amount of self-selection to the group of terminal users. Only the personnel who were most interested in working on terminals became terminal users at the company. All terminal users had their eyes tested and were given spectacles if they needed them.

Research in Sweden into the visual-ergonomic aspects of VDT work must be encouraged and its scope extended beyond the work situation to cover leisure time. This type of study is becoming increasingly important because more and more people have personal computers in their homes, which increases the total time they spend in front of a screen. This study is germane to the current debate on the relationship between exposure to VDTs and miscarriages. Various kinds of psychological stress are to be found among people who work at a terminal for long periods each day, and the occurrence of health problems has also been proved. More interdisciplinary research work is urgently required, combining medical and behavioural science. Not until this work is completed can we reach any reliable conclusions on cause and effect. Preventive measures must be taken now on the basis of current findings.

Long versus short daily periods at display terminals

We compared responses from people who use terminals for less then thirty minutes a day and people who use terminals for more than two hours a day.

The psychosocial work environment

Only indices that measured psychosocial work environment factors gave significant differences. People using the terminal for a short time each day feel that they get more information and they are more satisfied with the information they receive about events at their workplace. Thus there is a risk that long periods spent at a terminal each day make it more difficult for personnel to assimilate both formal and informal information at their workplace. Women and older employees who use a terminal for short periods each day are more satisfied with the opportunities they have to use their own knowledge and qualifications in their work. People who spend short periods each day at the terminal more often have a positive attitude to promotion. This group has a greater incidence of people who have duties and positions that provide the prospect of promotion and/or people who are 'pursuing a career'.

The effects of the terminal system

We were able to identify a number of significant differences between people who worked a short period and people who work a longer period at the terminal each day. The pattern varies, with both negative and positive attitudes present in each of the groups. People who work at a terminal for a long period each day are not so anxious about making mistakes and feel that the terminal system reduces the risk of error and gives them more opportunities to correct errors. The system has built-in checks, and these positive factors are of more benefit to the people who spend many hours a day at a terminal. Being used to terminal work also plays a part here. On the other hand, at company 1 (the Postal Giro Service), people were anxious about making mistakes, particularly in the Data Processing Department. The consequences of errors were enormous and would affect the operations of the company at large as well as the customers. The personnel in the Data Processing Department felt that they had too much responsibility.

Under the heading of the buffer role, the differences between the groups may be said to be that people who work for a long time at the terminal every day and who feel that heavy demands are made upon them from various quarters are mainly to be found in the Customer Service and Systems Analysis Departments. Personnel in the Customer Service Department have direct contact with the general public. The kind of work to be done on recently introduced types of insurance policy may be one reason why demands on the Customer Service Department have increased since the questionnaire survey was carried out. A comparison at the departmental level of the indices measuring the effects of the terminal system and the computer system in general shows that the Eastern Market Division, which includes the Customer Service Department, considers more than personnel in other departments of the company that the terminal system and the computer system have made the buffer role more prominent in their work. In the Systems Analysis Department, conflicting demands and the intermediate position appear to be built into the professional role of the personnel and were not directly affected by the terminal and computer systems.

These findings may be compared with the findings in company 1 (the Postal Giro Service) where the picture was rather different. There, the Data Processing Department could be described as the organization's 'buffer' department. The—often conflicting—demands coming from different quarters in the company appear to be concentrated or accentuated in the Data Processing Department. The personnel in this department thought that the EDP system had helped to increase the demands placed upon them and helped create a more pronounced buffer role situation. The demands were mainly from other departments or personnel groups in the company whose work was directly dependent upon the Data Processing Department. But demands also came from other associated groups. The opposing conditions in the operations side and the data processing side could be explained in part by the fact that the Data Processing Department was isolated and that there were major differences in the composition of personnel in these departments. Most of the Data Processing staff were young men, while more women worked in the operations departments. A company's data processing department can easily become a 'buffer department' in the organization.

Rather unexpectedly, the staff that worked a short time each day at the terminal thought that the introduction of terminals and computerization had given them worse working hours in terms of their opportunities to combine their working lives and their private lives. One interpretation is that personnel who spend a short daily period at the terminal work in a different way than those who work for long daily periods at a terminal. One of the departments whose personnel work long periods every day at the terminal is the Customer Service Department, where work stops at a fixed time every day, when the telephone switchboard closes. There are probably more people in supervisory and more qualified positions in the group that spends a short time every day at the terminal. Analyses show that people in supervisory jobs complain more than other groups about difficulties in reconciling their working lives and private/family lives.

The people who spend long periods every day at a terminal feel that there has been an improvement in the ergonomic environment. More attention should have been given to this group as against people who spend a short time every day at the terminal.

Health

The responses to the question on health show that personnel who spend a short time each day at a terminal complained less frequently of physical pain during working hours. The personnel who spend a long time at a terminal every day are to be found more often in the group with a lower level of education (the nine years of compulsory schooling) and in the group with a higher education (university graduates/higher education qualifications). This difference is not to be found in a comparison of people with qualifications higher or lower than the Swedish school-leaving certificate. The people who spend the longest time at a terminal are to be found in Systems Analysis and the Eastern Marketing Division.

The personnel who use terminals for a short period every day work mainly on general investigative work and the internal processing of customers' cases. The people who use the terminal for many hours a day are mainly the data entry personnel and personnel in the Customer Service Department.

Long-term experience of display terminal use

Comparisons were made between personnel with varying amounts of experience of terminal work. We compared people within the group of terminal users who have worked for less than one year and more than one month with people with more than one year's experience of terminal work.

The psychosocial work environment

The predominant findings on the psychosocial work environment are that people with short-term experience of working on VDTs have a more positive attitude to the system. They feel that they are exposed to fewer demands in their work from various quarters; they are not so troubled by the buffer role, fewer of them find the pace of work too high, they more readily accept the amount of responsibility and attentiveness the work requires and are not bothered by the different work load factors, such as too uneven a work pace and the pace of work being too dependent on other people.

The responses from women show that women with long experience of terminal work have a more positive attitude about the different factors that affect job content. For example, more women in this group than in the group with little experience of terminal use feel that they have better opportunities to utilize their knowledge and skill in their work, and do not feel that the work is too unskilled for them. This result may be compared with the breakdown into terminal and non-terminal users for the indices of *interesting work—meaningfulness* and *interesting work—human resources*. Women with more experience of terminal work are also more satisified with the office premises and the ergonomic design of their work station. To some extent, these results may be attributed to the gradual introduction of improvements at the company (see Chapter 9).

Democracy at work

Attitudes to different psychosocial work environment factors are often reflected in the responses to our questions on different aspects of democracy. People with less experience of using terminals were more satisfied with the amount of influence they had on scheduling breaks, holidays and working hours.

Health

Finally, people with less experience of using a terminal suffer less frequently from pain in their leisure time. The younger group in particular were troubled by pain.

Age and sex

As mentioned earlier, age and sex were kept constant in all comparisons in this chapter on results in order to avoid erroneous conclusions on the demonstrated differences. The age spread is more or less the same in the group of terminal users for groups with varying lengths of experience of terminal use. This is important to note since the older people in the total group reported more than the younger group that they suffered pain. The sex spread varies between the groups. Most of the people who have used terminals for a long time were women (70 per cent women and 30 per cent men), while there was an even spread (50 per cent men and 50 per cent women) in the group with less experience of terminal work. Notwithstanding the difference in sex representation in the groups, the tables show no difference between the respective groups of terminal users. The differences in health between people with more and people with less experience of using VDTs at work cannot be attributed to age and/or sex.

Display terminals and customer contacts

The psychosocial work environment

In the total group, personnel who frequently or sometimes had contact with the general public in their work showed a higher degree of satisfaction with their work tasks. The variables of the perceived meaningfulness of the work, the opportunity of applying personal experience and knowledge and the amount of job autonomy apply here. These results are reflected in the fact that personnel place a higher value on intrinsic aspects of the work, that is to say, having varied work, having influence on decisions that affect their own work and their own working conditions and having contact with other people in their work. But at the same time, personnel with more contact with the general public report a greater degree of dissatisfaction with their pay. They have also been affected somewhat more by the buffer role at work, and more often felt that the pace of work was too hectic and the work load too heavy.

The picture is somewhat different in the terminal user group. The strongly positive aspects of the job reported by people who often or sometimes have contact with the general public do not recur in these results. On the other hand, the less positive attitudes towards the buffer role and the more negative attitude towards the workplace and the workload remain. Thus display terminal users with more contact with the general public are more troubled by the varied and often contradictory demands related to their work role and more frequently feel that the work pace is too high, the work is too stressed and too high demands are placed upon them in terms of efficiency. Moreover, the results show that terminal users who have more contact with the general public are also more troubled by an irregular pace of work and the need for attentiveness/responsibility than terminal users who rarely or never have contact with the general public.

Effects of the display terminal and computer system

What, then, do people think about the display terminal system and the computer system in general in terms of the work environment, democracy-related issues, physio-ergonomic conditions, etc.? Among all personnel, people who are more exposed to the general public consider that the display terminal and computer systems have enhanced the buffer role, and that their pay structure was somewhat worse. They stated more frequently that computerization and the introduction of VDTs had hardly helped reduce the risk of physical and psychological problems. To a certain extent, these results are also to be found among terminal users. People with contact with the general public, then, consider that the demands made of them from various quarters are greater and that their pay levels have worsened. Moreover, terminal users with more contact with the general public feel that the computer system has reduced the influence they have on their own work and their influence in the company in general.

In the future, more and more people will be using display terminals in their work. The findings of studies made in this field point to several factors which must be considered in display terminal work: psychosocial work environment factors, the degree of democratization of the workplace, ergonomic and health issues (see Figures 2 and 4). The factors affecting these components at a workplace equipped with display terminals include the length of experience of terminal work, the length of daily terminal use and on-the-job contact with people such as customers and the general public. Other research on ergonomic factors (Gunnarsson and Söderberg, 1979) has shown that terminal-intensive work causes eye, neck and back problems. The findings of the RAM Project indicate that from a psychosocial viewpoint it is important to avoid intensive, continuous work on visual display units.

Secretaries and other female employees: A comparison

In this section, two different employee groups with different work tasks are compared. The duties that are often included in a secretarial job are set out in Chapter 6 together with changes in the work brought about by the introduction of word and text processing technology. Here, secretaries are compared with other female personnel involved in insurance policy administration with regard to the different 'blocks' of problems: psychosocial aspects of the work environment; democracy at the workplace; effects of the display terminal system and computerization in general; and health problems.

The results showed that the secretaries were significantly more dissatisfied with the following aspects of their work and working conditions: how interesting the work was; the amount of interesting work in terms of the utilization of human resources; job autonomy; opportunities for contact with co-workers; cooperation with co-workers. They were also more dissatisfied with promotion prospects; democracy and influence at work; their ability to influence working hours and free time. The secretaries were more dissatisfied with those areas in their work which are usually seen as being vital to job satisfaction. This is confirmed by the fact that their health was worst (statistically reliable): they suffered more frequently from stomach trouble and nervous problems.

However, they had a significantly more positive attitude (statistically reliable) about the new technology, i.e., the effects of the display terminal system on work pace, the attentiveness and responsibility factor, effects on physical and psychological problems related to the new technology—both the display terminal system and the computer system in general in the company. Thus they appeared to regard technology as an opportunity to free themselves from unsuitable professional roles.

They had used word and text processing systems for several years, this company being one of the first to introduce this technology. At many workplaces, this professional role is still in the development phase. Both the opportunities and the risks in the future design of professional roles become apparent when new technology is introduced (see Chapter 6). For a more detailed study of the results, see *The Working Roles of Secretaries and Word Processing Systems* (Bradley, Bergström and Sundberg, 1984, Ch. 4).

Inter-departmental comparisons

Table 5.13 provides a picture of the ways that the display terminal system changed working conditions in the company as a whole. The table contains the mean values and spread for the indices that measure the effects of the display terminal system (in some cases, the terminal and computer system in general) for all departments in the company. These tables may be read in different ways. For example, the departments can be compared index-by-index. Here the broad structure only will be considered. There are three departments in particular that differ from the others, namely the Data Processing Department, Company and Industrial Insurance and the Eastern Marketing Division (which includes the Customer Service Department).

Introducing the display terminal system in the Data Processing Department increased the interesting content of the work (TISJA), and increased the risks of errors (TFEL), while personal influence on one's working conditions improved slightly (TINFLAD), and wage levels (applies even to the effects of computerization in general) also improved (TDLÖN). On the other hand, the attentiveness and responsibility factor (TUPPA) was greater in the Data Processing Department than in other departments.

The Company and Industrial Insurance Department pointed to a number of factors that had worsened as a result of the introduction of the display terminal system. At the time of these investigations the so-called FIF system was being installed. The degree of interesting and autonomous work decreased (TISJA) and the risk of errors in work increased (TFEL). Influence both on the individual's work (TINFLAD) and in the company in general (TNFLFD) decreased. Finally, the personnel in this department reported more often than people in other departments that the introduction of display terminals involved certain health risks (THÄLSOR).

Reactions were mixed in the Eastern Marketing Division. The display terminal system had reduced the risk of errors (TFEL), and had improved working hours (TDARBTID) and ergonomic conditions (TERGO). But the buffer role was stronger (TGBUFF), as a result of the introduction of the display terminal system and the computer system in general, and the attentiveness and responsibility factor (TUPPA) had increased more than in other departments.

Some of the results have highlighted a number of physio-ergonomic

Table 5.13 Mean values (M) and spread (S) for indices which measure the effects of the terminal system and the data processing system in general

Index	Index Abbr.	Group	Accounts	Systems	Private	Personal	Company and Industrial	Occupational	Easten Marketing Division	Others	Total	
Interesting work/job autonomy – T	TISJA	3.04	3.13	3.52	3.08	3.08	2.76	3.15	3.08	3.00	3.08	M
		.12	.83	.69	.48	.35	.77	.55	.48	.00	.57	s
Errors in work – T	TFEL	2.98	3.05	3.25	3.03	3.04	2.75	2.94	3.33	3.00	3.04	M
		.08	.57	.68	.30	.47	.73	.32	.72	.00	.55	s
Contact opportunities with co-workers – T	TKONTSAM	3.00	2.91	3.17	3.00	3.10	3.02	3.07	3.05	3.00	3.05	M
		.00	.59	.46	.19	.30	.26	.15	.20	.00	.28	s
Intermediate position/buffer role – T/D	TDBUFF	2.86	2.76	2.83	2.94	2.89	2.88	2.81	2.54	3.00	2.85	M
		.31	.60	.40	.17	.24	.20	.31	.49	.00	.32	s
Influence – work democracy – T	TINFLAD	3.00	2.98	3.12	2.85	3.00	2.67	2.92	3.05	3.00	2.94	M
		.00	.13	.45	.46	.21	.52	.28	.33	.00	.38	s
Influence – company democracy – T	TINFLFD	3.00	3.00	3.08	3.02	2.96	2.82	3.04	3.02	3.00	2.98	M
		.00	.00	.18	.11	.31	.51	.09	.07	.00	.26	s
Irregular work pace – T	TOJA	3.00	2.67	2.67	2.73	2.77	2.67	3.00	2.81	3.00	2.78	M
		.00	.43	.43	.55	.35	.53	.20	.77	.00	.48	s
Attentiveness – responsibility – T	TUPPA	3.00	2.72	2.64	2.84	2.75	2.78	2.92	2.57	3.00	2.78	M
		.00	.44	.69	.37	.50	.53	.19	.53	.00	.47	s
Wages – T/D	TDLÖN	3.08	3.06	3.30	3.02	3.05	2.99	3.12	2.95	3.03	3.07	M
		.28	.19	.43	.07	.32	.32	.36	.22	.09	.30	s
Working hours – T/D	TDARBTID	2.91	2.83	2.81	2.80	2.86	2.84	2.66	3.11	3.15	2.87	M
		.60	.69	.58	.76	.72	.61	.60	.71	.76	.68	s
Interest in education – requirements – T	TUTIK	3.35	2.78	2.75	3.09	3.12	3.25	3.35	3.19	3.59	3.14	M
		.90	.67	.84	.80	.60	.70	.55	.54	.92	.74	s
Ergonomic factors – T	TERGO	3.00	2.96	3.08	3.06	3.07	2.93	3.03	3.20	3.00	3.05	M
		.00	.30	.26	.30	.23	.33	.18	.31	.00	.27	s
Temperature/air – T	TTEMPL	2.96	2.65	2.95	2.95	2.97	2.97	2.92	2.90	3.00	2.94	M
		.14	.75	.21	.54	.19	.18	.28	.64	.00	.39	s
Health hazards – T	THÄLSOR	2.58	2.44	2.41	2.78	2.45	2.24	2.54	2.48	2.64	2.50	M
		.67	.88	.70	.67	.95	.68	.80	.68	.71	.76	s
	Antal ca	17	12	30	40	51	32	14	21	14	172–234	

problems. Many of these problems can be attributed to the open-plan office. Some results show that people find factors such as overheard telephone conversations to be disturbing. Finding a peaceful atmosphere in which to work and difficulties in concentrating were also mentioned as problems. The open-plan office principle was not wholly compatible with one of the standard ergonomic requirements for VDTs, namely lighting. The company tried to put the problem right by setting up screens round the display terminal work stations. The company had invested heavily in the general ergonomic and physical aspects of the work environment. In many respects no expense had been spared on the company's physical environment, and experts in ergonomic environmental conditions had always been involved both in the initial phases and at a later stage, when improvements were made.

Some of the physio-ergonomic problems may be attributable to the phenomena described in the following letter from a reader of my book, *Occupational Roles and the Quality of Life* (Bradley, 1979):

> I have been a safety representative for three years. We have lodged complaints about the height of counters and about chairs and tables, all of which are concrete factors, and there has been a response to our demands. But the psychosocial environmental factors are taboo. I ask myself why, and believe that I have the answer. When we had replaced all the chairs and counters with furniture of the right height and dimensions, where could we lay the blame when we do not dare to acknowledge what the real problem is? Now, when we can no longer blame our bad backs or, now the fitted carpeting has been replaced, our respiratory problems, what do we do with our anxiety and fears? Our fears of the 'buffer role', the role that so many of us find ourselves in but do not dare to tackle; our feeling of inadequacy, and going home in the evening having failed to realize all our good intentions. I am in the fortunate position of being able to voice these thoughts, but what about all my fellow workers who do not dare to speak up because they don't dare to be like me, a troublesome complainer who is never satisfied, and who is looked upon as slightly odd. 'You shouldn't take everything so seriously and think so much'; better to say nothing, better to go home and ignore the situation, be content, be happy and endure the stomach aches rather than acknowledge the way things really are.

In general, the evaluation of the impact of technical change is a complex area of research. Other changes were taking place in the organization both before and at the same time as the terminals and computers were installed. For example, there were several mergers and, because systems were merged and new, more exacting work routines were introduced; people with routine duties and terminal users were more affected than other personnel. However, we have tried as much as possible to minimize the difficulties in reaching conclusions. Firstly, the general strategy has been to use different data collection and data analysis methods (see Chapter 3), using the analyses of interviews, internal documentation and quantitative methods. Our findings from the series of interviews, for instance, formed the point of departure for the specification of problem areas and subsequently formed the basis of the 'interpretation' of the results. Secondly, in applying the quantitative method, we used comparisons between different groups that differ from each other in their relationship to the technology, and also used questions where the people interviewed were asked about the effects of the introduction of the display terminals and data processing system. Finally, the results were referred to the joint reference group, where the conclusions were discussed in detail with people who had extensive knowledge of the conditions and developments in the organization.

4. A summary of the results

Comparisons were made between display terminal users and non-users. Relevant background factors were held constant, and only statistically reliable differences are presented below. In general, terminal users were more dissatisfied with some aspects of the psychosocial work environment, including issues of democracy at the workplace. They experienced more frustration in their work. They saw their work pace as being too irregular, too hectic at certain times and affected to a great extent by other people. Terminal users were less positive than non-users about promotion.

The work environment factors such as work pace, manning levels and efficiency requirements are the most dominant areas in which terminal users expressed a higher level of dissatisfaction than non-users. This picture does not remain constant when the groups are broken down by sex. For both sexes, the 'terminal-sensitive' areas of the work revolve around job content and in particular the amount of job autonomy and how meaningful the work content is. However, female terminal users are more positive about the content of their work than are non-users. The opposite is true for men. Moreover, efforts to achieve quality at work often collide with the process of communication. This was the case at company 1 in the project (the Postal Giro Service), where the content of women's work was 'depleted'. The relationship between sexual equality and computerization is seen as a vital issue for the future (see Chapters 9 and 10).

Terminal users point to several ways that computers have changed their work environment: there were greater demands placed upon them in their work (buffer role accentuated), the pace of work was more irregular and the attentiveness, accuracy and responsibility factor was greater. The introduction of the new technology made it possible to accelerate the rate at which work was performed and material processed. At the same time, the amount of contact with customers, etc. increased. However, it appears to be extremely difficult, if not impossible, to translate this increase in the rate at which cases are processed into an enhanced quality of personal contact, or into more time available for customers. As the work pace at the display terminal speeds up, it tends to set the pace for other functions. In addition, as demands from the employees' surroundings increase, the individual may gradually assimilate them and in the end they will become internalized.

The responses on health issues reveal that terminal users suffer more than non-users from general pains and headaches. Among the younger group, terminal users reported periodic eye troubles and even nervous complaints during their leisure time. Other workers suffered less frequently from these ailments.

A comparison of groups with different lengths of experience of work at the display terminal reveals that it is the terminal users who spend less time in front of the display terminals who have the most positive attitude. However, it is difficult to reach any conclusion about the effects of long-term use from these results. Among women, it is those with longer experience who have more positive attitudes to the work content itself.

Groups that spend different lengths of time each day at a display terminal were also compared. One risk of long daily use is the deleterious effect on the user's ability to assimilate both formal and informal information at work. But those who work longer hours worry less about making mistakes and have more opportunities to correct errors. They feel that the new computer system has reduced the risk of errors in their work. Nevertheless, the intrinsic demands of their professional role (the buffer role) have increased.

These results are based on direct comparisons between mean values. Analyses of a number of stress factors reveal that personnel who work long hours at a display terminal more often feel that they receive too little information, that there is too little prospect of promotion and personal development and growth, and the requirement of training/further training is too small. This group more often feels imprisoned by the work situation and isolated from other people during working hours, and also feels that the physical loads on backs, muscles and eyes is too great. On the other hand, people with longer hours of daily display terminal use feel more confident about their work and have a greater sense of job security.

Comparisons were also made between personnel with different degrees of contact with the general public. In general, personnel with regular or occasional contact with the general public showed a higher degree of job satisfaction, but were also more discontented with their wage levels and with the effects of the buffer role. Comparisons between the different groups of terminal users showed that the more positive effects of their work situation vanished, while the less positive attitudes to the buffer role, the work pace and workload remained.

In responding to a direct question about the impact on work environment of the display terminals system and the data processing system in general, along with matters of democracy, etc., the personnel group with more contact with the general public felt that the buffer role was accentuated and that their salary conditions were somewhat worsened. Terminal users with more contact with the public also felt that computers had reduced the influence they had over their own work and in the company as a whole. (See the previous section for comparisons between different departments, different professional functions, and between people with and without customer contact.) The results show that from the psychosocial perspective of the work environment and working life, *it is vitally important to avoid intense and continuous work at display terminals.*

These results are summarized in Table 5.14, which is a compilation of the information presented in the original table of results. As before, only statistically reliable mean differences are presented. The level of significance is $P<0.05$. When $P = 0.05$, the difference has been stated in brackets. Plus and minus signs are used to indicate the direction of differences, the plus sign (+) indicating that the group has a higher mean value than the comparison group and the minus sign (−) that the mean value is lower for this group than for the comparison group. Higher mean values indicate a more positive attitude or that the individual's circumstances are positive (in the case of three indices, Weighting—external, Weighting—internal and Weighting—development, a high mean value signifies that this factor is not considered to be important). The table may be read horizontally or vertically. Some headings in Table 5.14 have been abbreviated. The full text of columns 4 and 7 is: Has frequent or occasional contact with the public/seldom or never has contact with the public. This presentation of results was mainly concerned with comparisons between different groups, e.g., terminal users and non-terminal users. It may also be of interest to see which components in the work environment (indices) did not give significant differences between the groups.

The comparison between men and women and between younger and older personnel shows that computerization has an effect on equality-related issues, mainly equality between the sexes (see also the next section). Comparisons on the breakdown by sex and age have been included in Tables 5.2–5.12 and in the general table to make these results easier to evaluate. The tables show, for example, that the difference between terminal users and non-terminal users is often not attributable to the difference in sex.

5. Female and male personnel: Comparisons

The earlier tables in this chapter kept age and sex as constants in the analyses. Tables 5.2–5.12 show significant differences between the total groups of men and women. Let us now examine these differences more closely. The differences between the sexes were analyzed for the various problem areas covered by the project's theoretical model, namely:

- psychosocial aspects of the work environment (14 indices)
- democracy at the workplace (5 indices)
- the effects of the display terminal system and computerization in general (14 indices)
- health problems (11 indices).

The multiple classification analysis (MCA) method was used in these analyses. The following problem areas are analyzed:

- How are the demonstrated differences between men and women affected by a number of individual variables (age, professional position, income, working hours) that can explain the demonstrated differences between the sexes? The selection of background variables is based on earlier analyses.
- Which of the independent variables in the MCA analyses contribute most to explaining the various dependent variables (such as the psychosocial aspects of the work environment, workplace democracy, the effects of the display terminal system, computerization in general, health problems)? This estimate is based on Beta and Eta weightings in Tables 5.15–5.17.

Appendix 1 contains a description of the content of each of the indices. A list of indices related to the theoretical models in Chapter 3 is to be found in Figure 4 in this chapter. This formed the basis on which analyses were made of any differences between the sexes in terms of:

- psychosocial work environment factors (S)
- democracy at the workplace (DEM)
- physio-ergonomic work environment factors (F)
- the effects of the display terminal system and the computer system in general (T and D)
- health problems (HA/HF)

Table 5.14 An overview

The group marked + has higher mean values than the other group (marked −)

		TOTAL				VDU USERS				WOMEN			MEN			YOUNGER			OLDER STAFF		
		Women Men	Younger Older	Non-VDT users VDT users	Contact often or sometimes Contact seldom or never	Use VDT <30 mins/day Use VDT >2 hrs/day	Used VDT 1 month–1 year Used VDT >1 year	Contact often or sometimes Contact seldom or never	Use query system Use dialogue system	Non-VDT users VDT users	Use VDT <30 mins/day Use VDT >2 hrs/day	Used VDT 1 month–1 year Used VDT >1 year	Non VDT users VDT users	Use VDT <30 mins/day Use VDT >2 hrs/day	Used VDT 1 month–1 year Used VDT >1 year	Non-VDT users VDT users	Use VDT <30 mins/day Use VDT >2 hrs/day	Used VDT 1 month–1 year Used VDT >1 year	Non-VDT users VDT users	Use VDT <30 mins/day Use VDT >2 hrs/day	Used VDT 1 month–1 year Used VDT >1 year
Interesting work – meaningful	SIME	− +			+ −					− +											
Interesting work – human resources	SIMR	− +			+ −					(− +)		− +	+ −	(+ −)						(+ −)	
Job autonomy	SSJA	− +			+ −					− +			+ −								
Contact opportunities at work	SKONT	− +																			
Cooperation with co-workers	SSAM																				
Salaries	SLÖN		− +		− +								+ −								
Promotion – positive	SBEFP		− +	+ −		+ −				+ −	+ −						+ −				
Promotion – negative	SBEFN																			+ −	
Promotion prospects	SBEFM	− +					− +														
Information	SINFO					+ −					+ −			+ −			+ −			(+ −)	
Intermediate position/buffer role	SBUFF	− +	− +		− +		+ −	− +										+ −			
Work pace/work load	SARBT			+ −	− +		+ −	− +				+ −	+ −					+ −	+ −	+ −	
Irregular work pace/work load	SOJA			+ −			+ −	− +							+ −				+ −		
Job status	SAANS	− +	− +									− +									
Attentiveness/responsibility	SUPPA	− +					+ −	− +				+ −						+ −			
Efficiency	SEFF																				
Company democracy	DEMF	− +	− +	+ −									+ −			+ −					
Company democracy – preferences	DEMFÖ		− +																		
Work/leisure time democracy	DEMAL	− +		+ −			+ −		− +						+ −			+ −			
Work/leisure time democracy – wishes	DEMALÖ	− +		+ −									+ −					+ −	+ −		
Work premises	FA											− +									
Temperature/air	FT	− +						− +											+ −		
Ergonomic factors	FE								+ −			− +									
Odours	FL	− +				− +													+ −		
Interesting work/job autonomy – T	TISJA																				
Errors in work – T	TFEL					− +					− +									− +	
Contact opportunities with co-workers – T	TKONTSAM															− +				− +	

Intermediate position/buffer role – T/D	TDBUFF			+ −	− +	+ −		− +		+ −	+ −		+ −			+ −	+ −		+ −		
Influence – work democracy – T	TINFLAD																				
Influence – company democracy – T	TINFLFD																				
Irregular work pace – T	TOJA			+ −						+ −			+ −			+ −			+ −		
Attentiveness – responsibility – T	TUPPA			+ −						+ −			+ −			+ −			+ −		
Wages – T/D	TDLÖN				− +			− +							+ −			(+ −)			
Working hours – T/D	TDARBTID					− +					− +									− +	
Interest in education – requirements – T	TUTIK			+ −		+ −				+ −									+ −		
Ergonomic factors – T	TERGO					− +					− +									(− +)	
Temperature/air – T	TTEMPL	− +		+ −						+ −					(− +)	+ −					
Health hazards – T	THÄLSOR															+ −					
Interesting work and job autonomy – D	DISJA	− +																			− +
Errors in work – D	DEFEL					− +															
Contact opportunities – D	DKONT				− +																− +
Cooperation – D	DSAM									− +											− +
Influence – work democracy – D	DINFLAD	− +						− +												− +	
Influence company democracy – D	DINFLFD		+ −					− +		+ −	(− +)										
Work pace/work load – D	DATAB	− +		+ −							+ −			− +		+ −			+ −		
Interest in education – requirements – D	DUTIK	+ −																			
Ergonomic factors – D	DERGO						(− +)														
Temperature/air – D	DTEMPL	− +							− +												
Health hazards – D	DHÄLSOR				− +																
Nervous complaints – A	HAN	− +																+ −			
Eye complaints – A	HAÖ	− +														+ −					
General pain – A	HAV	− +	+ −	+ −		+ −								+ −		+ −					
Stomach complaints – S	HAM																				
Headaches, etc. – A	HAH	− +		+ −												+ −					
Nervous complaints – F	HFN	− +					(+ −)									+ −		+ −			
Eye complaints – F	HFÖ	− +																			
General pain – F	HFV	− +	+ −	+ −			+ −									+ −		+ −			
Stomach complaints – F	HAM																				
Headaches, etc. – F	HFH	− +		+ −													− +				
Sleeping problems – F	HFS	− +	+ −															+ −			
Effects of work on leisure time/family life	AF	+ −	− +										− +						− +		
Leisure time activities – D	DFSYM											+ −		− +						− +	
Contact – D	DFKM	− +																			
Weighting – external	VY	− +																			
Weighting – internal	VI	+ −			− +																
Weighting – development	VU	+ −	− +		− +																
Interest in promotion	VB		+ −	− +			+ −					+ −	− +						− +		+ −

Table 5.15 Psychosocial work environment factors, work democracy, physio-ergonomic work environment factors: a comparison between the total groups of women and men (unadjusted) and with various background data taken into consideration (adjusted)

INDEX		Unadjusted				Adjusted			
ABBREV.	INDEX	W	M	s-level	Beta	W	M	s-level	Beta
SIME	Interesting work – meaningful	4.17	4.39	<0.01	0.17	4.32	4.18		0.10
SIMR	Interesting work – human resources	3.35	3.80	<0.01	0.23	3.60	3.44		0.08
SSJA	Job autonomy	3.50	4.09	<0.01	0.37	3.65	3.86		0.13
SKONT	Contact opportunities at work	3.46	4.12	<0.01	0.37	3.54	3.99	<0.01	0.25
SSAM	Cooperation with co-workers	3.92	4.13		0.12	3.98	4.05		0.04
SLÖN	Salaries	2.95	2.93		0.01	3.11	2.71	<0.01	0.24
SBEFP	Promotion – positive	4.18	4.17		0.03	4.21	4.16		0.05
SBEFN	Promotion – negative	2.33	2.33		0.00	2.34	2.31		0.02
SBEFM	Promotion prospects	3.03	3.24	<0.05	0.14	3.13	3.06		0.05
SINFO	Information	3.33	3.52		0.10	3.42	3.40		0.01
SBUFF	Intermediate position/buffer role	3.51	3.77	<0.05	0.16	3.48	3.78	<0.05	0.18
SARBT	Work pace/work load	2.88	3.09		0.13	2.91	3.05		0.09
SOJA	Irregular work pace/work load	2.73	2.94		0.12	2.70	2.98	<0.05	0.17
SAANS	Job status	3.45	3.74	<0.01	0.17	3.62	3.53		0.05
SUPPA	Attentiveness/responsibility	2.88	3.18	<0.05	0.16	2.93	3.09		0.08
SEFF	Efficiency	3.59	3.59		0.00	3.53	3.66		0.08
DEMA	Work democracy	2.58	2.92	<0.01	0.29	2.68	2.78		0.08
DEMF	Company democracy	1.15	1.40	<0.01	0.38	1.22	1.29		0.11
DEMFÖ	Company democracy – wishes	1.64	1.69		0.07	1.64	1.68		0.06
DEMAL	Work/leisure time democracy	2.34	2.77	<0.01	0.45	2.45	2.62	<0.05	0.18
DEMALÖ	Work/leisure time democracy – wishes	1.88	1.97	<0.01	0.22	1.90	1.94		0.10
FA	Work premises	2.66	2.75		0.04	2.66	2.76		0.04
FT	Temperature/air	2.24	3.17	<0.01	0.36	2.32	3.06	<0.01	0.29
FE	Ergonomic factors	4.04	4.08		0.03	4.04	4.09		0.05
FL	Odours	3.37	3.92	<0.01	0.26	3.38	3.89	<0.01	0.24

Level of significance:
$p < 0.05$

Table 5.16 The effects of the terminal system and the data processing system in general: a comparison between the total groups of women and men (unadjusted) and with various background data taken into consideration (adjusted)

INDEX ABBREV.	INDEX	Unadjusted W	Unadjusted M	Unadjusted s-level	Unadjusted Beta	Adjusted W	Adjusted M	Adjusted s-level	Adjusted Beta
TISJA	Interesting work/job autonomy – T	3.03	3.14		0.09	3.08	3.08		0.01
TFEL	Errors in work – T	3.00	3.06		0.06	2.97	3.10		0.12
TKONTSAM	Contact opportunities with co-workers – T	3.04	3.07		0.05	3.05	3.05		0.01
TDBUFF	Intermediate position/buffer role – T/D	2.85	2.85		0.01	2.84	2.86		0.04
TINFLAD	Influence – work democracy – T	2.92	2.97		0.07	2.93	2.95		0.02
TINFLFD	Influence – company democracy – T	2.96	3.00		0.08	2.97	2.99		0.04
TOJA	Irregular work pace – T	2.75	2.79		0.04	2.79	2.75		0.04
TUPPA	Attentiveness – responsibility – T	2.78	2.77		0.01	2.93	2.72		0.12
TDLÖN	Wages – T/D	3.05	3.10		0.08	3.04	3.10		0.10
TDARBTID	Working hours – T/D	2.93	2.79		0.10	2.92	2.80		0.09
TUTIK	Interest in education – requirements – T	3.13	3.13		0.00	3.13	3.13		0.00
TERGO	Ergonomic factors – T	3.02	3.07		0.10	3.01	3.08		0.13
TTEMPL	Temperature/air – T	2.88	3.01	<0.05	0.15	2.90	2.99		0.11
THÄLSOR	Health hazards – T	2.46	2.52		0.04	2.44	2.55		0.07
DISJA	Interesting work and job autonomy – D	2.91	3.18	<0.01	0.26	2.95	3.14	<0.05	0.19
DEFEL	Errors in work – D	2.94	3.04		0.10	2.91	3.08		0.15
DKONT	Contact opportunities – D	3.02	3.05		0.06	3.04	3.02		0.02
DSAM	Cooperation – D	3.02	3.10		0.14	3.03	3.09		0.10
DINFLAD	Influence – work democracy – D	2.83	2.98	<0.05	0.15	2.88	2.92		0.04
DINFLFD	Influence – company democracy – D	2.94	2.99		0.09	2.96	2.96		0.00
DATAB	Work pace/work load – D	2.59	2.80	<0.01	0.22	2.58	2.80	<0.05	0.23
DUTIK	Interest in education – requirements – D	3.25	2.96	<0.01	0.20	3.25	2.96	<0.05	0.21
DERGO	Ergonomic factors – D	3.05	3.03		0.02	3.04	3.04		0.00
DTEMPL	Temperature/air – D	2.90	3.02	<0.05	0.14	2.94	2.98		0.06
DHÄLSOR	Health hazards – D	2.52	2.56		0.03	2.53	2.56		0.02

Level of significance:
$p < 0.05$

Table 5.17 Health problems: the effects of work on leisure time/family life and the effects of computerization on opportunities for leisure activities: weighting of work environment factors: a comparison between the total groups of women and men (unadjusted) and with various background data taken into consideration (adjusted)

INDEX ABBREV.	INDEX	Unadjusted W	Unadjusted M	Unadjusted s-level	Unadjusted Beta	Adjusted W	Adjusted M	Adjusted s-level	Adjusted Beta
HAN	Nervous complaints – A	3.52	3.87	<0.01	0.26	3.55	3.80	<0.05	0.18
HAÖ	Eye complaints – A	3.43	3.98	<0.01	0.30	3.44	3.93	<0.01	0.26
HAV	General pain – A	3.54	4.29	<0.01	0.40	3.64	4.15	<0.01	0.27
HAM	Stomach complaints – S	4.28	4.41		0.09	4.33	4.36		0.02
HAH	Headaches, etc. – A	3.69	4.13	<0.01	0.23	3.73	4.06	<0.05	0.17
HFN	Nervous complaints – F	3.51	3.84	<0.01	0.24	3.54	3.80	<0.05	0.19
HFÖ	Eye complaints – F	3.54	4.04	<0.01	0.29	3.56	4.01	<0.01	0.26
HFV	General pain – F	3.81	4.38	<0.01	0.34	3.87	4.29	<0.01	0.25
HFM	Stomach complaints – F	4.34	4.48		0.11	4.38	4.42		0.03
HFH	Headaches, etc. – F	3.81	4.26	<0.01	0.24	3.84	4.21	<0.05	0.19
HFS	Sleeping problems – F	4.16	4.40	<0.05	0.15	4.21	4.32		0.07
AF	Effects of work on leisure time/family life	4.21	3.91	<0.05	0.16	4.04	4.13		0.05
DFSYM	Leisure time activities – D	2.99	2.99		0.02	2.98	3.00		0.09
DFKM	Contact – D	2.99	3.03	<0.05	0.16	2.96	3.05	<0.01	0.33
VY	Weighting – external	2.30	2.76	<0.01	0.34	2.42	2.56		0.10
VI	Weighting – internal	2.28	2.03	<0.01	0.18	2.17	2.18		0.01
VU	Weighting – development	2.96	2.63	<0.01	0.20	2.94	2.66	<0.05	0.18
VB	Interest in promotion	2.93	3.03		0.05	2.85	3.14	<0.05	0.14

Level of significance:
$p < 0.05$

– the effect of work on leisure time/family life (AF) and the effects of computerization on leisure time activities (DF)
– weighting of different factors in the work environment (D).

The following is a brief account of the results of the MCA analyses as a whole. Some of the results on health were published previously in RAM report no. 11, *Computers and Their Psychosocial Impact, Especially with Regard to Women*, and an article in *Occupational Health Nursing Journal* (Bradley, 1983a, 1983b).

Psychosocial work environment factors

The tables on the different indices used to measure the psychosocial work environment (S-indices) show that women personnel are consistently less satisfied with the various aspects of the psychosocial work environment than are men. In all of the indices where there are significant differences between the sexes, women show a lower mean value, that is to say they are less satisfied/more dissatisfied. This applies to the following indices:

	Significant difference disappears	Significant difference remains
Interesting work, meaningfulness (SIME)	x	
Interesting work, human resources (SIMR)	x	
Job autonomy (SSJA)	x	
Contact opportunities with co-workers (SKONT)		x
Promotion prospects (SBEFM)	x	
Intermediate position/buffer role (SBUFF)		x
Status of work (SAANS)	x	
Attentiveness and responsibility (SUPPA)	x	

The demonstrated sex differences disappear when the four background variables of promotion, working hours, age, and income are taken into account. On the other hand, the significant difference between men and women in opportunities for contact with co-workers remain when these four background variables are taken into account. Here, women are more dissatisfied, and the sex of the worker interviewed is a fairly important factor in explaining attitudes to the opportunity for contact with co-workers.

The significant difference for the index 'intermediate position/buffer role' also remains when these variables are taken into account. Women are more exposed than men to contradictory demands and are more troubled by finding themselves 'in the middle' in their jobs and in professional roles that involve dealing with conflicting demands.

In two cases, a difference arises between men and women when work supervision, working hours, age, and income are taken into account. Men are more dissatisfied with their wage levels. When these four variables are constant, the explanatory value of the 'sex' variable becomes more marked.

Democracy at the workplace

The results from the above analyses of psychosocial working environment factors are reflected in the indices that measure influence and democracy at the workplace (the DEM indices). Significant differences are found in the following four indices:

- work democracy (DEMA)
- company democracy (DEMF)
- work-leisure democracy (DEMAL)
- work/leisure democracy, preferences (DEMALÖ).

The differences demonstrated between men and women disappear for all but one of the indices when work supervision, working hours, age, and income are taken into account. Women are more discontented with the amount of influence they have on the scheduling of breaks, holidays and working hours (within the framework of the flexible working hours system) (DEMAL). However, there is a substantial drop in the explanatory value of the 'sex' variable (ETA = 0.45, BETA = 0.18). It seems that the salary level has the greatest explanatory value for various aspects of democracy and influence at the workplace.

Physio-ergonomic work environment factors

Women are more troubled than men by temperature and ventilation conditions (FT), odours and smoking at the workplace (FL). These differences remain when the variables of work supervision, working hours, age, and income are taken into account.

The effects of the display terminal system and the computer system in general

It is important to note that in the indices that measure the effects of the display terminal system (T-indices) there are no significant differences by sex regarding various aspects of the psychosocial work environment, democracy at the workplace, and ergonomic conditions. The same results are obtained when the four background variables are taken into account, i.e., no significant differences appear. The only difference demonstrated is that women more often contend that the display terminal system worsened the temperature and ventilation conditions at the workplace (TTEMPL). This difference disappears when the four background variables mentioned above are taken into account.

Women personnel say more often than men that computerization in general (D-indices) has caused the work situation to deteriorate. This applies to the following indices:

	Significant difference disappears	Significant difference remains
Interesting work/job autonomy – D(DISJA)		x
Influence/work democracy – D(DINFLAD)	x	
Workplace/workload – D(DATAB)		x
Temperature/air – D(DTEMPL)	x	
Educational interest/requirements – D(DUTIK)		x

The significant differences here remain for three of the indices when the four background variables of work supervision, working hours, age, and income are taken into account. Thus female and male personnel had similar perceptions of the effects of the new display terminal system. As to the perceived effects of computerization in general, women feel that computerization has lessened the variety in their jobs, the interesting job content, the oppportunity to apply their own ideas, knowledge and skills, the amount of job autonomy they have, as well as the status and prestige of their jobs (DISJA). Moreover, women more often than men feel that the computer system (computerization) in general has brought about an increase in the degree to which other people affect the pace of their work, irregular work pace, stress at work, and the degree to which machines and equipment dictate the pace of work (DATAD). Women also feel more often than men that computerization affects their own wish for more education and their interest in education, as well as the company's requirements for education and further education (DUTIK).

For the indices that measure the effects of computerization on influence at work and work democracy (DINFLAD) and the effects of the computer system on temperature and air (DTEMPL), the differences between men and women disappear when the four background variables mentioned above are taken into account.

Health

The following conclusions may be drawn from analyses of the health factors (HA and HF indices). There are significant differences between women and men for all health indices, with women reporting these problems more frequently. One exception is stomach complaints, for which there are no differences.

Five indices cover health problems during working hours:

Nervous complaints (HAN)
Eye complaints (HAÖ)
General pain (HAV)
Stomach complaints (HAM)
Headaches (HAH).

Six indices cover health problems experienced during leisure hours:

Nervous complaints (HAN)
Eye complaints (HAÖ)
General pain (HAV)
Stomach complaints (HAM)
Headaches (HAH)
Sleeping difficulties (HFS).

The significant differences remain when the background variables of supervision, working hours, age, and income are taken into account, with the exception of sleeping difficulties, for which the significant difference disappears. The explanatory value of the 'sex' variable drops for the indices 'nervous complaints' and 'general pain' when the four background variables of work supervision, working hours, age, and income are taken into account.

The influence of work on leisure time/family life and the effects of the computer system on the opportunity to pursue leisure activities

Women display a more positive attitude to the effects of work on leisure time/family life (AF). Women feel that their present position has very little effect on family life or leisure activities (overtime and unscheduled working hours, etc.). However, the difference disappears when the background variables of work supervision, working hours, age, and income are taken into account. On the other hand, women more often than men feel that computerization at the company has both directly and indirectly limited their opportunities to take part in home and family life, to associate with their friends, or to attend training courses or undertake further education, i.e., different opportunities for contact with other people outside working hours (DFKM). The 'sex' variable has a strong explanatory value here.

Some analyses of the effects of psychosocial working environment factors, influence and democracy at work and ergonomic conditions on health and on the effects of work on leisure time/family life are set out in Chapter 8. The interactions between computer technology, working life and leisure time/family life are analyzed in *The Secretary Role and Word Processing: Office Profession in Transition* (Bradley, Bergström and Sundberg, 1984) and in *The Role of the Engineer and the CAD System*, RAM report (forthcoming). Peter Martens has applied the RAM Project's theoretical model and measurements of indices to further analysis of the relationship between work and leisure time/family life (RAM report no. 12).

The weighting of different work environment factors

The people interviewed made an assessment of how important they felt different areas at work and their working conditions to be in terms of their well-being at work. Female personnel at the company place a higher value than men on external aspects, i.e., good working hours, a good physical working environment, a calm and even pace of work, and a job that can go hand in hand with one's family and leisure-time activities (VY). Male personnel place a higher value than the women on intrinsic aspects of work, i.e., having varied work, work that allows personal development, being able to exercise personal influence on decisions relating to one's own work and working conditions (VI). However, these two demonstrated differences in the evaluation of work and working conditions disappear when the background variables of work supervision, age, and income are taken into account.

A pattern of evaluation similar to the content of the index 'weighting—intrinsic' (VI) occurs in yet another index, 'weighting—development' (VU), which means that good opportunities for promotion, good opportunities for further training, and having a job with scope for personal development are considered to be important. Men place greater value than women on these aspects, and the difference remains when the control variables are taken into account. It appears that the age variable has the greatest explanatory value.

A significant difference between the sexes occurs in the index 'interest in promotion' (VB), when the background variables of work supervision, working hours, age, and income are taken into account. Men report the greatest interest in promotion; here too, the age variable has the greatest explanatory value.

The factors that people see as important in their work are often a reflection of objective conditions at work. Under certain circumstances people tend to place less importance on the areas where they feel they have less to gain and place greater importance on the areas where they gain more. Added to this are selection mechanisms that apply to the wishes and aspirations of the individual when he or she begins full employment. These phenomena are also interwoven with experiences at work and have been analyzed in RAM report no. 4, and in Bradley (1970, 1972).

A great deal may be read from the previous tables and the description of the results. The results may be interpreted with the aid of human engineering theories, with psychological and sociological theories on sexual divisions in the community and differences in conditions of life and the roles and values of men and women (see Bradley, 1972, Ch. 10). The results indicate that a detailed study must be made of the demonstrated differences between the sexes and that these differences can be traced back to differences in background conditions, for example, data on occupation, family circumstances, conditions in working life, etc.

It is also important that account be taken of any general differences by sex that have been demonstrated. In the computerized society of the future it is important to emphasize any differences between the sexes in perspectives on and perceptions of the world we live in, both when formulating future goals and when developing strategies to attain these goals. It is also important that the experience, values, and perspectives of women are considered in the models of reality that form the basis of systems development and which are expressed in research in the field of computer science.

6. Follow-up interviews

It is five years since I wrote the report, *The Work Environment and Display Terminals*, and it is seven years since we completed the different interview and questionnaire phases in the fieldwork. What has happened since then? In the summer of 1985 I met a former staff representative of the Swedish National Union of Insurance Company Employees who was a member of the joint reference group in our research project, a physiotherapist who was occasionally co-opted into the reference group, and the personnel director and one of the heads of the Data Processing Department.

Interview with Kerstin Enstedt, former staff representative, Swedish National Union of Insurance Company Employees

Gunilla: What is your general impression of the introduction of computer technology at Trygg-Hansa?

Kerstin: I think that the organizational changes in Trygg-Hansa were more significant than the process of computerization. There have been many organizational changes—that is very common in business. The latest change is an attempt to merge the Claims and Marketing Departments to create a sense of unity among people working on insurance policies. This system is already in effect in our regional and local offices, but we still have to introduce it here at the head office. By doing this we aim to give personnel a better overview of the business—it will integrate the insurance and claims operations.

Gunilla: So the principles applied in earlier reorganizations have been applied here too, where a single employee follows a case from beginning to end, rather than splitting the work between several people. There were a great many changes at Trygg-Hansa from 1977 onwards. The open-plan office was completed, new technology was introduced, new organizational structures were tested, and there was a spate of mergers. What happened after that?

Kerstin: Well, not much has happened since then. The process of change has been a very gradual one.

Gunilla: Have you now reached a fairly stable level of technology?

Kerstin: That is hard to say. I recall that in the middle of the 1970s our union discussed the total number of employees that there should be in the insurance industry. All our fears about the future proved to be groundless. We cannot say that it is technology that has caused us to lose jobs. In fact the number of employees at Trygg-Hansa has increased. We were seriously worried about government reports about employment. In Trygg-Hansa we tried to derive some comfort from the fact that the workforce's age profile showed that everybody would be pensioned off in time. But in fact we have taken on more staff in recent years.

Gunilla: Have you increased the volume of production and services offered by Trygg-Hansa?

Kerstin: Possibly, in the case of life insurance. Since the pace of change is so slow now, no staff groups have suffered. Things are pretty quiet here.

Gunilla: Has the company's international independence been a factor here?

Kerstin: If you have money, you can be more considerate about rationalization.

Gunilla: What are your comments on the events of 1977 to 1980?

Kerstin: We have never really been afraid of computerization here at Trygg-Hansa. When it comes to industrial democracy, we attempted to construct systems that guarantee employee influence. Trygg-Hansa also tried to start up a major training programme. What people complained about in 1977, which was the point I had to put to the reference group, was the rate at which these display terminals were to be introduced. The people scheduled to get them last thought that this was unfair. But we had to make sure that the telephone links would work, we were waiting for new machines to be supplied, etc.

Gunilla: Among other things, we found a number of indications of an increase in stress among display terminal users. This was expressed in the form of physical and psychological complaints. The pace of work had been forced up. What is the picture today? Is a process of adaptation in progress?
Kerstin: We have had no complaints for quite a while.
Gunilla: People who spent a long time at the display terminal every day suffered more from ailments and were more stressed. What is the current situation?
Kerstin: We have avoided routines that involve people in long periods at the display terminals. The work routines now include both paperwork and work at the display terminal. When systems are built, we design jobs so that they comprise a long chain of work tasks. For example, one employee would deal with an insurance case from start to finish. We tried to use terminals to give people the opportunity to do other things as well. In both the Life Insurance and Property Insurance Departments we tried to stick to the maxim 'one person, one job'.
Gunilla: Our investigations did not show that job content had been depleted, but, as I said, there was more stress.
Kerstin: Our physiotherapist in the company health service may be able to tell you something about that. We tried to locate the display terminals so that we could measure the volume of work and see how response times were affected, so that we could prevent long response times. When the computerization programme was changed in 1977, we had a breathing space, a chance to think over the way we wanted things to be. That was useful. Technology should serve as an aid, and that is not always the case. We shelved the idea of a typing pool using word and text processing technology, and we have secretarial services in each department. Rationalization can be done efficiently if properly thought through and carried out at the proper pace. I also think that it is profitable to invest in training.
Gunilla: What happened as a result of the trade union measures that were introduced?
Kerstin: The 'Co-determination and Advanced Technology' programme is still in progress and we have taken some further action. We present reports to our national conferences. Our request to the TCO (Swedish Confederation of Salaried Employees) for a temporary halt to computerization was very controversial. Our 1982 national conference dropped that demand because it was considered too unrealistic. As things were in 1980, there were fears about employment levels in the industry. Reports from the government commission on the impact of computerization caused a lot of concern in the banking and insurance industries. Two years later, things had calmed down. Now we want to negotiate agreements on technology, but management

is not interested. I think the co-determination agreement and the work environment agreement give us enough power to pursue an aggressive policy on these issues. But you have to learn to use these powers, and that is something we have managed to do. There is a separate data group in our staff association that monitors the present computerization programme and possible future developments. We are beginning to talk about office automation, for example, the introduction of the paperless office. There was talk of this several years ago. But I think computerization and the introduction of display terminals have actually increased the volume of paperwork. I have seen a lot of this sort of thing before, so I take all this quite calmly. Our fears have often proved groundless. Not least, researchers have painted a picture of change as something to be frightened of. People have become used to the idea of resisting it. But things haven't turned out remotely the way they were predicted.

Gunilla: Are the prognoses unrealistic: does one see what is technically possible? Or do we in a little country like Sweden have the instruments we need to nip any undesirable trends in the bud?

Kerstin: It may be that painting these scenarios prompts the response from some trade unions that they will never accept such situations, and that has an effect on future trends.

Gunilla: What was the result of your trade union action?

Kerstin: Well, we sent that letter requesting a temporary stop. And the response was, 'Oh, really?' They wondered if we really meant what we had written.

Gunilla: What are your views on people's fears of what the future holds, and what they expect from the future?

Kerstin: The issues related to computerization were very prominent when I was chairman of the union staff association. The process of computerization and its effects took up all our time and interest. I neglected a lot of other issues. I argued these issues in the Group board and in the management group. It may be that all the work we put into these matters has made the whole system work, and then perhaps we can leave these issues and spend more time on other matters. Computerization hasn't posed any problems since 1979. This has also been noticeable in the activities of the staff association when we attempt to take up these matters. If people are worried about these things, then they go to the meetings and discuss them there. The monitoring of computer-related issues has now been delegated to a special computer group.

Interview with Anita Linder, physiotherapist

Gunilla: What are your comments on our findings?

Anita: At the time you carried out your investigation, many people

The use of personal computers is becoming more widespread.

Photo: Nils-Johan Norenlind/TIOFOTO

were still spending a lot of time at display terminals every day. They were doing only one part of a job. The claims adjuster, for example, made a settlement, but the claims adjuster did not key in this information himself at the terminal; the documentation was sent in batches to people who keyed in the information for two hours at a stretch. This category of personnel no longer exists today. Everybody has a display terminal. The claims adjusters key in their own claims and this lets them spread the terminal work over a whole day. I try to teach people to do this. Terminal work has become more integrated with other work tasks. This change was made only recently in the case of claims adjusters, who were initially against it.

Gunilla: Even when we were here, there were attempts to reorganize the handling of insurance business so that one employee could follow a case from beginning to end. Is there any group of staff at all that uses the display terminals a lot nowadays?

Anita: Yes, the programmers in the Data Processing Department. I have just completed a questionnaire survey of that department. All eleven people sitting by the windows had eye disorders. They were young people who did not need reading glasses—if they did, they had been provided with them. They reported doing between four and seven hours work at the VDT. The older this group becomes, the more difficult it is for them to manage this work. As a programmer, you use this terminal as an aid and you spend a lot of time at the terminal. Programmers must constantly check their work on the screen. I believe they are bothered so little by this work because they are quite young, most of them are men who take a lot of exercise in their leisure time, and they also find their work very interesting. They don't make complaints. People who were a little older complained of eye problems and stress. Computers are being used more and more in other occupations too. There has been a break now, when computers were integrated into other work tasks. Now a number of work tasks are beginning to disappear. People are now beginning to say: 'Good heavens, we are beginning to use computers more and more.' The claims adjusters are one example of people who are using computers more and more. They may not spend so much time looking at the screen, since they concentrate more on the keyboard. But they see that VDT work is on the increase. However, the picture may be quite different in five years.

Gunilla: What is the current situation in the Customer Service Department?

Anita: They used to share terminals, they had group terminals. Having to get up and move around is a good thing in itself, but people often had to queue to use the display terminal and the waiting periods were long. The waiting periods have been shortened by introducing ancillary computers and microcomputers. Colour screens have been introduced and if, for example, an insurance policy has run out, it is displayed in a particular colour.

Gunilla: The Customer Service Department staff found it difficult to use the telephone and the terminal at the same time.

Anita: Let's not forget that when you were here, the display terminal was still a very new aid. Everyone was against them at first. Now people have a very positive attitude. They see that they couldn't do the work without a terminal. There is chaos when the computer system breaks down. And the terminal programming in the Customer Service Department has improved. They used to have to go through many stages to retrieve the information they wanted. And many causes of stress have now been removed. Some people complain, of course. But there have hardly been any complaints from the Customer Service Department for quite a while now.

Gunilla: What about word and text processing?

Anita: Many of our secretaries have begun to use word and text processing software. This is the group with the most sedentary work, but they don't look at the screen so much. Most of their initial difficulties were running-in problems. There were secretarial services on each floor, with most people using word processors and the rest using ordinary electronic typewriters. Some of the directors have their own secretaries, which was unthinkable seven years ago. One of the secretarial service units is a typing pool. We haven't received more complaints from the secretaries than from other staff. The staff in our Office Printing Department sit still a lot. We are testing a rotation system at the graphic display terminals, but there are a lot of complaints about the hours people have to spend sitting in front of a VDT.

Gunilla: Is it true that people make greater demands on themselves and push themselves harder when they use equipment that increases their efficiency? Or do people think that now they can do so much more than before, they can take things a little easier?

Anita: People don't complain of stress. But neck, arm and shoulder problems begin to emerge—often caused by too rapid repetitive movements. There are fewer lower back complaints now. People don't relate display terminals to stress, but say that other things make them stressed. Somehow, people have accepted the terminal, and they feel it cannot be removed. But the display terminal is certainly an indirect cause of a lot of neck and arm problems. The amount of stress caused by the open-plan office has largely been resolved by putting people whose jobs require attentiveness into rooms of their own.

Gunilla: What about computerization, stress, and the managerial position?

Anita: We can see that in the departments in this house where we have a lot of patients there is often a supervisor who does not fit in well with the team. So the management function is a very important one.

Gunilla: Do you still use the display terminal handbook?

Anita: As basic material, yes, but there is a lot more recent information published. Office managers have *The Office Handbook*. This book contains rules for VDT work. And we also run our own internal courses for all Trygg-Hansa safety representatives and work supervisors.

Gunilla: What are your expectations and/or fears about computerization and people's working conditions in the future?

Anita: I think that the amount of time spent at display terminals will increase. Even though people say the opposite is true, I think we can already see that it is on the increase. So I believe that people need to become far more aware of their bodies and their souls. We physiotherapists must teach more about relaxation, breathing, and how to take breaks. We must make it clear to the work supervisor that these are important factors, so that they don't suspect their staff of shirking. We shall have to create recreation opportunities in the work environment. I would like to begin with the staff in the Data Processing Department because this is a group that is beginning to have problems. I think that VDT work will increase, although it might take a few years. The Customer Service Department will have the same work situation as it does now. The sales staff will be using computers. The claims adjusters and insurance policy groups will be using computers more and more. The managers have begun to use IBM PCs, but so far only a few people have them.

Interview with Torbjörn Bergström, personnel director

Gunilla: What are your views of the main findings of our investigation and what is the situation today?

Torbjörn: That is a good question. There has been a process of adaptation, and people have become used to this type of workplace. We have also equipped the work stations with tables, chairs, and lighting that are suitable for use with display terminals. Work stations used to be converted in a rather slapdash fashion. But we have staff who still suffer from eye complaints, who ask for filters and different types of screen. This is also a generation question. People who have used a card register all their lives thought that the change was too great. Many young people, who were part of the data age at school, have a different attitude. I belong to the generation which thinks that this development is rather intrusive. But we have expertise, a person in our Office Administration Department who works on job design, and two physiotherapists who advise us on ergonomic questions.

Gunilla: Are any of your staff working solely on data entry today?

Torbjörn: As far as I know, data entry is a natural part of our staff's work and no-one works purely on data entry. We have a Data Registration Department linked to the Data Processing Department, which deals with the work that is not done on display terminals, for instance, work from the Wages Office and the Finance Department. But even these departments are being equipped with display terminals.

Gunilla: What has happened regarding efforts to make changes in management when computer technology was introduced? At that time, you had management recruitment problems.

Torbjörn: Management positions are not attractive. Sales managers do not have very good prospects of making a lot of money. But this evens out in the long term. I see some problems for managers in the office administration area. These are not company policy problems, or personnel problems, but socio-political problems. People do not want to expose themselves to the stress and tension that is inherent in managerial positions when there are virtually no financial rewards. The effects of the taxation system help to dampen the urge for personal advancement.

Gunilla: When the Personnel Department was computerized, there was an experiment in which managers exchanged positions with one another.

Torbjörn: They managed to do this very well in the Personnel Department but we have not succeeded in other areas. We would like to have more job rotation, but we are faced with a 'better the devil you know than the devil you don't' attitude. Managers do not want to lose assistants who are good. We are doing quite a lot of intensive work on managerial and personnel development. But a lot of very talented men and women today say that they don't want to become managers because their families are more valuable to them than a career in management and their net pay increase will not compensate for the problems they will have to contend with. Many people are prudent enough not to apply for managerial positions.

Gunilla: What organizational changes have there been in recent years?

Torbjörn: We have reorganized into divisions. In 1980 we introduced autonomous profit-making departments. Decision-making powers have been moved lower down the organization. Staff management, for example, has been placed closer to the people it concerns.

Gunilla: Have you used a local computer system to support this organizational change?

Torbjörn: We are still working at group level with mainframe computers and the introduction of display terminals. But we do have some minicomputer systems, on which I refer you to Roger Fastesson, one of the heads of the Data Processing Department.

Gunilla: Which has been the more important, reorganization or computerization?

Torbjörn: The introduction of a division-based structure has been a more important change, while terminalization was a more natural aspect of the work. Setting up a new structure of divisions, with new work tasks, new managers, and new terms of reference has had a greater effect. Then there are fewer of us and we are working in larger office areas. The number of separate offices has increased. But we are no longer facing major changes such as mergers and moving into an open-plan office.

Gunilla: Although the number of staff has been reduced, have you maintained the same production volume?

Torbjörn: Yes, there has actually been a considerable increase in our production volume.

Gunilla: How accurate were the forecasts made in the government enquiry on employment?

Torbjörn: After your investigation, rationalization brought about a steady decrease in the number of employees up to 1983, but since then our staff has expanded. There were about 6000 people employed in the companies that merged to form Trygg-Hansa, and today we have a staff of 3800.

[At this point Gudrun Widén, who has worked at Trygg-Hansa for thirty-three years and is currently head of selection and recruitment, joined the interview.]

Gudrun: The number of staff diminished by a little over 100 people a year from 1977 to 1981. I refer here to the entire Trygg-Hansa Group, including both the office and sales sides. The size of the workforce then remained constant through 1982 and 1983. But throughout 1984 the Group has taken on more people, and now the total number of staff is on the increase.

Gunilla: How do you account for these fluctuations?

Gudrun: The major staff cutback of 1977 was caused by the Group's financial position. We were forced to make cuts in staff, staff costs being 70 per cent of our expenditure. This decision was taken at a high level in our organization. Then the situation changed and there was a fantastic boom in personal insurance. It was very easy to sell pension insurance policies from 1982 to 1983. Sales increased by 40–50 per cent.

Gunilla: Which occupations have been cut back, and which have increased?

Gudrun: Mainly typists' jobs in our typing pool, because of the introduction of modern word and text processing machines. We had 250 people in that personnel group, and this number has been halved. We also use standard letters a great deal now. It is difficult to measure the reduction. The introduction of display terminals has meant that many work tasks have been transferred from the head office to our field offices. There is an increase in staff on the sales side. We have 691 salesmen today. The figure was 600 at the end of the 1970s. We have not computerized work in the Personnel Department as much as in the rest of the company. Jobs now have a different profile. Risk management has been introduced, we now have graduate engineers on our staff, and the work has become more complex. Because of the increase in production volume, we are taking on more staff than are leaving.

Gunilla: Are the staff becoming better qualified?

Gudrun: There are enormous differences in terms of education. This is related to the current labour market situation. When we can employ people with academic qualifications for the same salary as less qualified staff, then we take on the better-qualified people. Many of the more simple routine jobs have gone. The answer to your question also depends on what you mean by a better-qualified staff? If you look at our educational requirements, then qualifications are higher. But in terms of job content, the old-fashioned insurance processing procedures were more qualified than the simple process of ticking off the basic information on a form or at a display terminal.

Gunilla: So you are employing better-qualified people for jobs that require lower qualifications?

Torbjörn: People with a better basic education, not better-qualified staff.

Gunilla: Will people feel overqualified for the work they are doing? The standard of education is rising and people are faced with job requirements that are not on a par with their level of education.
Torbjörn: This is also a question of the hopes people form while they are at school. For instance, people used to start here as trainee claims adjusters and over a five-year period learned the insurance business and how to make evaluations and deal with customers. Today, because of their high qualifications, people graduate to these jobs in a matter of months. But they haven't learned the business.
Gudrun: At the same time, increasingly complex data processing systems are being installed, which means we must have staff with qualifications that never used to be required. The work becomes more abstract. People used to have their jobs in their hands, so to speak. The introduction of this technology has a lot of negative aspects, but there are also considerable advantages.
Gunilla: What are the prospects for future technological change? Have we achieved a stable level of technology or a situation in which the level of technology will rise rapidly?
Torbjörn: Technical change is accelerating. The latest innovation is in sales. People handling complicated life insurance policies need to access computers. When visiting customers, our staff use the telephone line to link a computer to the mainframe computer. By keying in the family's profile, the date of birth and the customer's requirements, a tailor-made quote is displayed on the sales representative's screen. In our terms, this is a major step forward, involving as it does a sales staff of 600. The same is true of the development planned for our claims assessors.
Gunilla: Has insurance policy administration become so complex that you are compelled to introduce computer technology? Has our entire society become so complex that computerization has become essential? Perhaps one alternative is to reduce the level of complexity.
Torbjörn: Or have we realized the enormous potential of technology and made no efforts to limit the degree of complexity, because we have the resources to handle it? Technological advance is accelerating. Our pools of experience in risk assessment, in calculating premiums, in legal knowledge and the application of regulations will be stored in computer memories. But the ability to get along with people, the ability to listen and to relieve the effects of a disaster, such as a disability or a burnt-down home, can never be computerized.
Gunilla: How are we to apply and develop that sort of expertise?
Torbjörn: By acknowledging that we need both of these areas, and we should not forget the importance of human contact by concentrating on computerization. It is important to translate computer terminology to a language that expresses human consideration and is not sterile and mechanical. There is a risk that people who have grown up with

this technology will easily fall into the habit of speaking very curtly when passing on to the customer the information they have retrieved from the terminal. It is important that we discuss these risks and that we provide service training. This is one of the tasks of the Personnel Department. There are three components: money, machines, and people. Money is there and must yield interest, and this is very much an ideological issue. In the sphere of machines, technology is advancing with bewildering rapidity. But when we come to people, we must not forget to follow up on trends, on research and, not least, the viewpoints that you and your colleagues have put forward. If we consider money and machines/technology to be the only important factors and think that everything else can take care of itself, then we are making a mistake. It is people who should control the other two factors and not the other way round.

Gunilla: Perhaps the interest in the functions of the two halves of the human brain is related to this. We must develop both halves of our brains, both logic and emotion and creativity, etc. in a technological society. Developing the special qualities that humans have, for example, our judgment and emotional involvement, has become increasingly important. What are your expectations and fears as a personnel manager in terms of the future development of new technology?

Torbjörn: I would like to see a less static organization, an organization in which changing jobs, leaving a managerial position, or transferring from a supervisory position to specialist work is a more pleasant and less traumatic experience. I would like all this to become less status oriented, so that being replaced is not a sign of major failure. I would like to be able to take the initiative on making a move without seeing it as a failure. I would like to see people displaying more civil courage, giving unequivocal responses, even if they are negative responses, and I would like to see people standing for what they say even if it is to their disadvantage in the short term. I wish we could all help each other more to stop thinking in terms of our own territories, and to place greater weight on individual competence than 'historical' background and status, and get closer, reaching over the boundaries of category and sector to encourage and support each other.

Interview with Roger Fastesson, Data Processing Department

Gunilla: What has happened in the field of hardware and software over the last five years?

Roger: There has been an explosive increase in the number of display terminals. We reckon to have some 2000 terminals by the end of this year, i.e., one and a half employees per terminal. We have been talking about entering terminals in our accounts under fixtures and fittings just as we do with telephones today and not entering them as a separate cost item. There has also been an increase in the amount of paperwork generated by the Data Processing Department. We thought that equipping the departments with display terminals would reduce this flow of paperwork because people would be getting the information on screen. Today, eight tons of paper are printed every day. It was only five tons a few years ago. People talked about a paperless office, but we have given up that idea. But the amount of paperwork may lessen in the future. Many of our customers do not have terminals and many of our staff still want their information on paper. This is also a matter of the generation gap, and the average age of Trygg-Hansa staff is still fairly high. If you trust the information you get on screen, you don't need to have it on paper. Younger personnel are eager to use this equipment, while older staff are afraid to press the keys and are afraid of doing something wrong.

Gunilla: Is there any category of staff in your department that has increased or decreased, or have any new categories emerged?

Roger: The group of staff working on data entry in this department has decreased, but not as fast as we thought. The number of systems men and programmers—we have no clearly defined boundary between the two—has increased. The large number of consultants has decreased. On the operational side, new professional groups have been added with the job of preparing and supervising terminal work. This is a pure control panel job using a large number of display terminals to supervise and support over 2000 terminals. This is an important part of the work of the Data Processing Department; these are the people who are our contacts with the rest of the organization. These are the people you call on when something goes wrong.

Gunilla: How is the Data Processing Department organized?

Roger: The department is divided into three groups. I am the head of the section of the department that takes care of everything that is not to do with insurance—that is, economic administration, finance and staff. There is one group for property insurance and one for motor insurance. I am also responsible for word and text processing, database processing and methodological issues for the Systems Department.

Gunilla: What is your view of computerization and employment?

Roger: The major rationalizations that were carried out in the 1960s and 1970s were in the banking and insurance industries. Routine jobs were mechanized. Since that time, the only changes have been in technology, i.e., the VDT has replaced the printed form as a means of presenting information. Each staff member has a higher work turnover now, but there are no longer any major effects felt from the rationalization process. I do not think that the number of staff will be reduced much more.

Gunilla: What is the design of the decentralized computer system?

Roger: We have developed a number of fairly large systems to decentralize power to the regional offices. In Claims, for example, the main register has been decentralized, and digests of information and statistics may be retrieved centrally. In fact, we only have mainframe computers at our regional offices now, with so-called ordinary dumb terminals connected to them. Today, some people have combined work tasks, for example, their work involves both producing quotes and word processing, and they are being equipped with PCs.

Gunilla: Have there been any other important changes in data processing?

Roger: Looking at software, we used to build large systems for a single purpose. We have begun to think of alternative uses of the data that we have already stored, the statistics and marketing data, for example. The major insurance routines are computerized and now we are beginning to look at the Finance, Economic Administration and Personnel Departments. Fourth generation languages have speeded up our work too, and programmes can be developed more rapidly. Personal programming will also be introduced for 'semi-professionals' who are supported by the Data Processing Department.

Gunilla: Has computer technology been used to support any particular organizational change?

Roger: There is no good answer to that question. This direction in computerization was already established before we had our present type of organization. However, this type of organization has allowed more people to become involved in this work. People felt that they had a chance to influence developments.

Gunilla: What are your views on the 'time spent at the terminal' problem?

Roger: Some of my colleagues in the data processing field report eye complaints and similar problems. I do not really know what should be done about that problem. We do everything at the terminal—we programme, test and produce documentation. We could, in principle, spend the whole day at a display terminal. Fortunately, we produce some lists which give us the chance to move around a bit. Systems engineers and programmers spend more time than any other category at the terminals. We shall have to see how things go. The open-plan office is an advantage in this respect; it may have a slightly dampening effect on the intensity of display terminal work. Personal contacts are easier in open-plan offices. Soon, only those of us working in the data processing field will have monochrome screens, and not colour screens. It was not until the last three years that every programmer was given his/her own display terminal. Now we have some pools equipped with terminals for testing and similar work. These pools allow access to several terminals at the same time.

Gunilla: Has anything happened in recent years in terms of the organization of systems development work?

Roger: The project form has worked well. An example is the claims project where one user is seconded to the project. The SIS – RAS* models work and are logical. But not enough attention is paid to the demand specification and people are too slapdash in their programming. The emphasis of the work should be at a later stage in the process. We also have a design for the roles in a systems development project and we have gained more expertise in project work.

Gunilla: What are your views on 'stress and terminal work'?

Roger: When I look around, the factors I notice are stress and response times. We were told at a seminar that response times should be 0.2 sec. Other sources say that you should have longer response times to allow you to think through what you are doing. It depends on what work you are doing.

Gunilla: Should you have a variable response time?

Roger: Yes, we should have that. If you are sitting opposite a customer, you want to have a particular response time.

Gunilla: Is my suggestion technically feasible?

Roger: It may be difficult to arrange. But we also have to deal with different work loads during the day, which we are trying to handle by increasing our computer power. Stress is a product of sitting passively. You used to have to go to a shelf to collect documents. Now, you do not have any influence.

Gunilla: Hopes and fears about the future?

Roger: Let's put it this way, new categories of people will be affected. Managers and administrative staff have got off fairly lightly so far. Now, managers will be spending more time on data processing, word processing, calculations and follow-up. This trend is heralded by office automation. The routine jobs are already done. We will now have to use intelligent terminals. Many more people will be doing a lot of their everyday work at a display terminal, PC or suchlike. Many people will be doing their own typing and producing their own documents.

Gunilla: Do you think that rationalization at a higher level will take place faster in Sweden than in other countries, for example in the USA?

Roger: We have actually come a long way in Sweden today. But we are a small country compared with the USA. In the Data Processing Department, we do the same type of work and the same amount of

---

* The Swedish Institute for Systems Development Guidelines for Administrative Development.

work. We will increasingly link up with different companies and their databases, and access and purchase information from the Post Office Administration, for example. The risk lies in having too much communication of this kind. But there are authorities in our country whose job it is to monitor that area. The National Data Inspection Board do their job well.

6.

An electronics company: the changing role of secretaries using word processing systems

1. Introduction

This chapter is a summary of a sub-study in the RAM: 2 Project. Its purpose was to describe and analyze the effects of the introduction of microcomputers when a group of office workers was equipped with a word processing system. The survey group consisted of female secretaries at Philips Electronikindustrier AB.

This sub-study is an in-depth follow-up of the research previously carried out in the RAM: 1 Project described in Chapters 4 and 5. The RAM: 2 Project is divided into three problem areas. On the basis of this division, study was also made of the nature of the work, its characteristics, and work organization in the company:

- computers and the communications structure at work;
- computers, leadership/cooperation and the roles of management/colleague;
- computers—working life—the conditions of life outside work.

More details on the background and purpose of this study are to be found in Chapter 3; the complete findings from the investigation are presented in *The Working Roles of Secretaries and Word Processing Systems* (Bradley, Bergström and Sundberg, 1984). The results can be generalized to apply to a larger group, namely female office staff, as the work tasks of the group under study correspond, by and large, to the general job profile of office staff.

The group under study was defined as all secretaries in the D-division (defence electronics). Everyone in the group used microcomputer-based word processing equipment practically every day, and at the time of the investigation had their own equipment; 48 per cent used the equipment between four and six hours a day and 24 per cent used it for more than six hours a day.

The sub-project on the role of the secretary: the project group included (left to right) Margareta Stenhjelm, Cecilia Bergström, Lilleba Sundberg, and Gunilla Bradley.

In Bradley, Bergström and Sundberg (1984) we discussed in greater detail the implications of word processing, its design and its present and future use. One chapter describes the organizational structure of the company, the background data for the group under study and the methods used. The study also includes a review of handbooks for secretaries produced in the post-war decades. These sources present a detailed profile of the secretary as a professional, well-qualified woman. However, this profile has been undermined in recent decades, the profession of secretary has come to be seen as an extension of the traditional female role and secretarial work has not been given a particularly high status. This is apparent from the more recent sources, including the books, analyses, papers and articles on these issues included in the review of literature.

2. Changes in the content of secretarial work

What are the duties of a secretary?

The duties that the secretaries at Philips Elektronikindustrier AB report that they perform during a working day are precisely the tasks that they consider have increased since the introduction of a word processing system, namely typing, booking trips and hotels and arranging meetings and conferences.

What changes in the objective work content occurred when word processing technology was introduced?

40 per cent of the people interviewed report an increase in the amount of typing work they do. This indicates that the volume of written material has increased. Long documents are no longer sent to a typing agency, for example. 60 per cent of the people interviewed state that they do more editing work. There has been some increase in booking trips and hotels, arranging meetings and conferences and preparing tables and illustrations. Both increases and reductions in the coordinating function were reported.

The increase in the volume of typing work is attributable to the computer being used as a means of improving efficiency. An appropriate use of human resources will benefit both the employees and the organization itself. Regrettably, this often leads to a situation where there is no time over for personal contact at work, such as with customers, or time for the further training that new technology requires (see Chapter 5). Instead, more and more of the same type of work is added to an individual's work load. The number of items to be processed and enquiries per unit of time has doubled, as has the volume of written work to be produced. At the same time, our own ambition levels and the demands we put on ourselves also increase. The blame for this cannot be laid at the door of technology itself; it is rather due to poorly thought out principles of administrative rationalization. If a new system is to be introduced, work tasks should be spread among different professional categories and levels, more or less at the same time. This provides an opportunity to reshape occupations which today are far too one-sided, among both senior administrative positions (executives and department heads), and positions and tasks involving more detailed work. So far, computer technology has been introduced mainly at the lower levels in companies.

The increase in editing work indicates a change in the content of secretarial work, and this means that secretaries will need further training. They now require training in subjects such as the use of language, editing and layout. Secretaries who want to see a reduction in the amount of typing and editing work they do put forward proposals related to job enlargement. They also mention organizing typing pools where simple typing work is concentrated in one unit either within or outside the company. This type of proposal can easily mask the need to examine job content and tasks as a work organization issue that should be discussed at the workplace.

The question of work organization is a matter of the designation of different positions and professional roles. It is obvious that there is some confusion about the terms used here to describe jobs. For example, there are rumours that people using word processors will be either 'upgraded to terminal operators' or 'downgraded to clerical positions'. This type of argument may be related to problems of one's own professional identity, rooted in an historical development linked to the application of new technology in office work. Another cause of confusion over job descriptions is that we are far too inclined to regard the organization as a purely static entity and not as one that is undergoing constant change.

Five aspects of change in job content occur when computer technology is introduced (see Chapters 1 and 9). Firstly, with regard to *psychosocial aspects of job content*, a large proportion of the respondents report this situation as unchanged. About half of those interviewed (52 per cent) stated that the amount of stimulation and tension in the work had increased, 44 per cent reported more variety in the job, and 40 per cent that they had greater discretion to apply their own knowledge and ideas.

A large proportion of the population also reported no change in the various aspects of *influence at work*. However, 60 per cent stated that their personal influence on the pace of work had increased. About one-third said that they had more influence on the planning of their work (32 per cent) and on the choice of working methods (36 per cent).

These results run completely counter to thc experience gained primarily from the first company in the RAM: 1 Project, a state-owned enterprise which used a batch-type data processing system. This was an example of centralized computing power—the heyday of the mainframe computer (see Chapter 4). The increase in influence on the pace, planning, and choice of working method in one's own job might indicate a possibility opened up by decentralized computer power in the form of word processing systems. Microcomputer-based equipment opens up new opportunities for job design. For example, the use of microcomputers means that the operator is no longer dependent upon telecommunications and a telenet for every task. Furthermore, other strategy options for introducing this technology open up (see Chapter 2).

In answering the questions on *work pace and dependence* on the equipment, as many as 64 per cent report that the need for attentiveness and accuracy increased with the introduction of word processing; 60 per cent also report an increase in the degree to which the work pace is controlled by the machine/equipment. The computer is de facto used as an instrument of rationalization. As a result, the work pace intensifies and the job becomes increasingly machine-controlled. *Extrinsic performance requirements* made possible by the new technology easily give rise to *intrinsic requirements* in the form of the demands we make upon ourselves both at work and in our private lives. Several references are made to this phenomenon in Chapter 5. *Our relationship to time* is, and will continue to be, one of the key questions for the future.

What will the future content of secretarial work be?

The possibilities and fears about the future impact of word processing concern both anxiety about the development of individual professional roles and the risk of becoming interchangeable. Faith in the future is needed too, and ideas on the positive development of this professional group. A number of examples are given of how to regain the former versatility in the profession irrespective of job designation: operator, assistant or clerk. An interest was shown in working on budgeting, planning and personnel administration.

3. Changes in communication patterns

Communication and computerization

The word 'communication' may have different connotations. Here, it is seen from the psychosocial perspective mainly linked to communication at work. The focus is on quantitative and qualitative interaction among people.

The process of change involving computer technology has a direct impact on communication, relative to the goal of the rationalization measures, or the purpose of the computer system. For example, when they use VDTs, people 'speak' directly to a computer, while the basic systems themselves are designed by people, for people (see Chapter 1).

Computer technology also has an indirect impact on communication. The type of technology and the way it is applied influence a number of factors in the work environment (organizational and structural design, the power hierarchy at the workplace, working hours). Certain phenomena in these areas of the work environment in their turn affect communication among people. Even though a number of the components necessary for a sense of belonging and cooperation to flourish at the workplace are now in place, we still bring to our professional group a wide range of positive experiences from our families, private lives and educational backgrounds which might complicate or even set at risk our relations at work. It is important—even from the work environmental viewpoint—to develop one's own personality in various ways (see also Chapter 9, section 3).

What factors at work and in the work situation help bring about good communication?

Secretaries use terms like 'openness', 'confidence', 'trust', 'respect' and 'information on work objectives' to describe the concept of quality in communication between people at the workplace. The factors they mention as helping to bring about good communication are usually associated with personal characteristics. In their answers, the people interviewed mentioned both contributing and preventive phenomena. They mention 'envy' but do not go into possible causes of envy, for example, the rewards system or people's status in the organization. They mention 'a feeling of independence' but do not raise the question of whether their own work tasks would accommodate independent action. It is a common phenomenon in all organizations to personalize problems of cooperation and to adopt a psychological approach to them. It is often difficult to analyze the causes related to objective, structural factors at different levels of analysis.

The obstacles mentioned are related both to the individual, such as reserve, fear and the inability both to give and take, and to factors in the work situation, for example, lack of time and too much work, the confusion of terms used to describe jobs, the issue of office space and an excessive number of managers.

Personal contact at work was wanted by 32 per cent in all contexts; 34 per cent in certain work tasks, for example, in two-way exchanges of information, customer contacts and contacts between departments in the company; 20 per cent when emotional factors are involved. As many as 16 per cent of the people interviewed stated that they did not want to have personal contact at their work in any circumstances.

On obstacles to personal contact, 44 per cent of the respondents mentioned psychosocial obstacles, namely work status and prestige considerations, and language barriers linked to the relations between professional groups; workload and stress; and personal causes which are indirectly attributable to work and the work situation.

The introduction of word processing has increased both work load and stress, and work load and stress both affect and constitute obstacles to personal contact at work. One of the physical obstacles to personal communication is that the word processing equipment curtails the staff's freedom to move about the office.

Our findings and experience with regard to communication indicate that it is important to distinguish between the concepts of seclusion and isolation at work, possibly in terms of voluntary and involuntary isolation (e.g., imposed by technology). At our workplaces, contacts between people and machines are on the increase, while there is a tendency towards less personal contact between people.

What changes in communication and contact have occurred with the introduction of word processing?

Three aspects of changes in communication were examined:

- *Quantitative aspects* (where contact has occurred, disappeared, increased, been dampened): there has been a considerable increase in contact with the boss, one's closest colleagues (technicians, engineers), and in customer contact; other types of contact are mostly reported as 'unchanged', but there are in general more contacts per time unit.
- *Means of contact* (face-to-face, by telephone, letter, telex, etc.): most of the people interviewed use word processing equipment which does not offer a communications option, and their responses reflect this situation; the option to link up microcomputer-based equipment via the telecommunications network completely changes the conditions for communication, from both technical and psychosocial viewpoints.
- *The content* of the reasons for contact at work: people's discussions have become increasingly technical (see the psychosocial obstacles for personal contact above); the language gap between secretaries and technicians/managers, partly caused by disparate educational backgrounds, is tending to change.

New, common work interests have emerged: 'You gradually begin to get a better understanding of computer terminology.' Does this mean that the real content of work and the results of one's work become more remote? Is there a tendency to focus more attention on the technology as a tool? It is difficult to discern a trend from the responses in the study, because the electronics industry produces the technology—both the hardware and software—in use here. Within the group of secretaries the word processing system has been a direct 'cause of contact'. The technology helped this group develop a stronger sense of professional identity and solidarity.

How can communication be improved?

The interview responses show that this question is associated with improvements in a technical sense, for example, a more rapid exchange of information, rather than the psychosocial aspects of communication.

What is the formal and informal contact network at the workplace?

The secretaries' meetings are an example of an informal contact network and the type of groups that form when this type of technology is introduced. There are different types of collective actions for women and for men. They take different forms in different countries, and at different levels in the emancipation of men and women.

4. Has the professional role changed?

A basic view of the working environment concept has been that there are different levels of the work environment. It has proved fruitful to use a concept that links the level of the individual and the level of the organization, for instance. The concept of the professional role is a building block, or rather the mortar between the social systems and the aggregate demands with which such systems confront their members as individuals. The use of this concept opens channels between socio-anthropological, sociological, and psychological theories, for example. Thus it links together phenomena such as cultural factors at the level of society, at the level of the organization, and at the level of the individual. At the same time, it is a term that we use in everyday language. We are familiar with terms such as the role of the parent, the management role, etc.

Most of the secretaries in the study work with a group of engineers. Over half the secretaries (52 per cent) feel that the introduction of word processing has changed their own professional role, and more than half the population sees the opportunity for professional advancement in the company in a positive light. They were clearly positive (72 per cent) about their own interest in further training.

In the secretaries' opinion no changes had occurred in the management role when word processing was introduced (72 per cent), nor in the interaction and cooperation between secretaries and management (76 per cent). The sex and professional roles are interwoven. According to the secretaries, the professional role is dominated by expectations, i.e., role expectations from the environment in which they practise their profession. In psychosocial literature, the term 'role' refers to the behaviour patterns related to both the duties and powers conferred by a certain social position, for instance, a profession. However, the individual can make use of the duties and rights in a position, i.e., adopt a role. Roles are not permanent. They are under constant change as conditions in working life and the community change, and they are under pressure from both intrinsic and extrinsic requirements and expectations. Role theory may be used to present and summarize some of the central results of our research, and role theory will also help clarify general discussions of computer technology and changes in working life.

5. *The effects on private life/family life and leisure time*

Both the opportunities for and interest in (percentages in brackets) most of the most common leisure time activities have changed with the introduction of new technology. This applies to training courses (40 and 56 per cent), home and family life (20 and 36 per cent), spending time with friends (20 and 24 per cent), reading (28 and 20 per cent), watching TV (24 and 20 per cent) and pursuing hobbies (24 and 20 per cent). These findings may be compared with the companies studied earlier in the RAM Project, where corresponding effects of computerization of working life on leisure time and its content were not demonstrated.

The research indicated an increase in collective action, but this is expressed here in the form of secretarial meetings, which are not part of trade union activities. There are many other reasons why female personnel show little interest in unions. Elsewhere, computerization is reported as having increased trade union activities. Electronics companies, however, are male-dominated workplaces.

There has been a marked increase in interest in training—an important prerequisite for the positive development of the professional role and new technology.

To questions on how one wants to spend one's time away from work, a larger percentage (60 per cent) of the population lists self-realization activities (studies, travel, cultural activities). Rewards expressed in 'self-realization' activities (stimulus, freedom, self-esteem, participation) weigh heavily in terms of gainful employment too. Social motives (contact with other people) are also fairly significant.

The time away from paid work (weekday evenings and holidays) is spent on pure relaxation and leisure by 32 per cent, on relaxation and leisure combined with extra work (housework in the evening, tidying up, washing, etc.) by 52 per cent, while extra work is performed by only 16 per cent of the group under study.

6. *Conclusions*

The study was monitored by a local joint reference group appointed by the safety committee in the company. The findings and observations were discussed with the reference group. The findings were also passed on to the secretaries' group, to the product managers in the sales organizations, and to key people in the company. During the spring of 1985, the results were circulated at conferences and discussion meetings about the future role of the secretary. These occasions provided an opportunity for discussing the following issues:

- What do you think should be done at the level of the society in relation to computerization—measures that are relevant to the work and working conditions of secretaries and clerical staff?
- What do you think the company and trade union should do when computer technology is introduced or action taken that affects the work situation of the secretaries and clerical staff?
- What action must be taken to promote the best use and development of your experience and knowledge?
- How can you personally tackle questions related to computerization and conditions of work?
- What developments are desirable for your professional group?
- What developments are desirable at your own workplace in terms of the shaping of future professional roles?
- How can you, as a member of a group at work, tackle issues intended to create better working conditions when computer technology is introduced?
- What should the standard requirements of a word processing system be?

7. Computerization and stress

1. What is stress?

Medically speaking, stress is our reaction to various kinds of irritation. Stress is a combination of defence and adjustment measures that is the organism's response to frustrations and other sources of irritation in its surroundings. In many cases this is a positive and appropriate reaction, while in other cases it is inappropriate or misdirected. The term 'stress' in everyday speech normally refers to these deleterious and harmful reactions. But stress can also be a positive reaction.

Hans Seyle, who worked from 1926 in Prague and later in Montreal, discovered psychological stress. The word 'stress', which is now a part of everyone's vocabulary, has been in use for about fifty years. Seyle developed theories on stress, using the concepts of overstimulation and understimulation. He also divided people into two types, the racehorse and the tortoise, with different tolerances of stress. Marianne Frankenhauser and Lennart Levi in Sweden are stress researchers of international repute.

Stress occurs in many different situations and it is often associated with working life. One typical stress situation is when individual achievement is balked and the satisfaction of individual needs is frustrated by demands and expectations from the organization, from managers, a group of co-workers or people outside the organization, for instance, customers. The lack of clearly stated requirements is also a common obstacle. Furthermore, technology and various combinations of technology can be a direct or indirect cause of stress reactions.

Stress occurs when an individual sees a situation as threatening and frustrating and cannot reduce his frustration by means of socially acceptable and generally appropriate behaviour. Thus individuals are often frustrated in their work when they cannot express the anxiety, anger, etc. they feel, and are compelled to subdue these feelings.

Psychosocial tensions or opposing pairs of factors can be found in working life, and these can give rise to stress (Levi, 1972). Such stress can cause discomfort in various parts of the body. We may feel heavy headed, have a stabbing sensation in the heart and pain in the midriff, tension in the jaw muscles, disturbed sleep and breathing difficulties (see Chapter 8 on health). Prolonged irritation can cause psychosomatic illnesses, the most common of which are stomach ulcers, high blood pressure and vascular diseases. These may be called the 'long-term effects' of a bad work environment.

Current research shows that prolonged continuous work at a VDT can cause problems in the neck, shoulders and back. These are caused mainly by poor seating posture and the fact that people spend far too long sitting and working at a VDT. However, it is reasonable to assume that these problems stem in part from the stress phenomena indirectly related to the use of technology, and attributable to the opposing pairs in Figure 7, Chapter 9. Neither one nor the other of these opposites, i.e., too much or too little of the different factors in the job, are good for the individual's mental well-being. These pairs have also been used in the RAM Project questionnaires (see below) to study changes in conditions of work brought about by computerization. (Most of the 400 questions are, however, of a different type.)

The inconsistencies between the needs of the individual, defined in one of the needs theories which are often discussed in human engineering, and the physical and psychosocial work environment, cause stress and may perhaps summarize the phenomena illustrated by these opposing pairs. Most problems are likely to derive from an individual being understimulated at work (right-hand column, Figure 7, Chapter 9). Large groups in both the blue- and white-collar sectors have jobs that are closely supervised, have some repetitive work operations where operators are confined by their equipment—conditions which lead to understimulation and a low degree of involvement in the job. The opposite condition—over-stimulation—should also be recognized as a work environment problem, and some tendencies towards overstimulation can be seen among white-collar workers. Rationalization, mainly in the form of computerization, eliminates the routine type of white-collar duties first. The people who carry out these tasks today will in the future increasingly have jobs and duties which are currently performed by people drawn from the upper echelons of white-collar workers, that is, specialists and the like. These positions will in their turn face restructuring when computer systems are introduced. In the white-collar area a number of closely supervised and fragmented jobs and work tasks have been created in the first stage of automation. However, these are likely to be eliminated in the next stage in the automation process. It is difficult to make long-term predictions of what changes will occur because measures introduced at different levels gradually become integrated in the total picture. In Sweden, training programmes and other activities are in progress, and these affect developments in a number of ways.

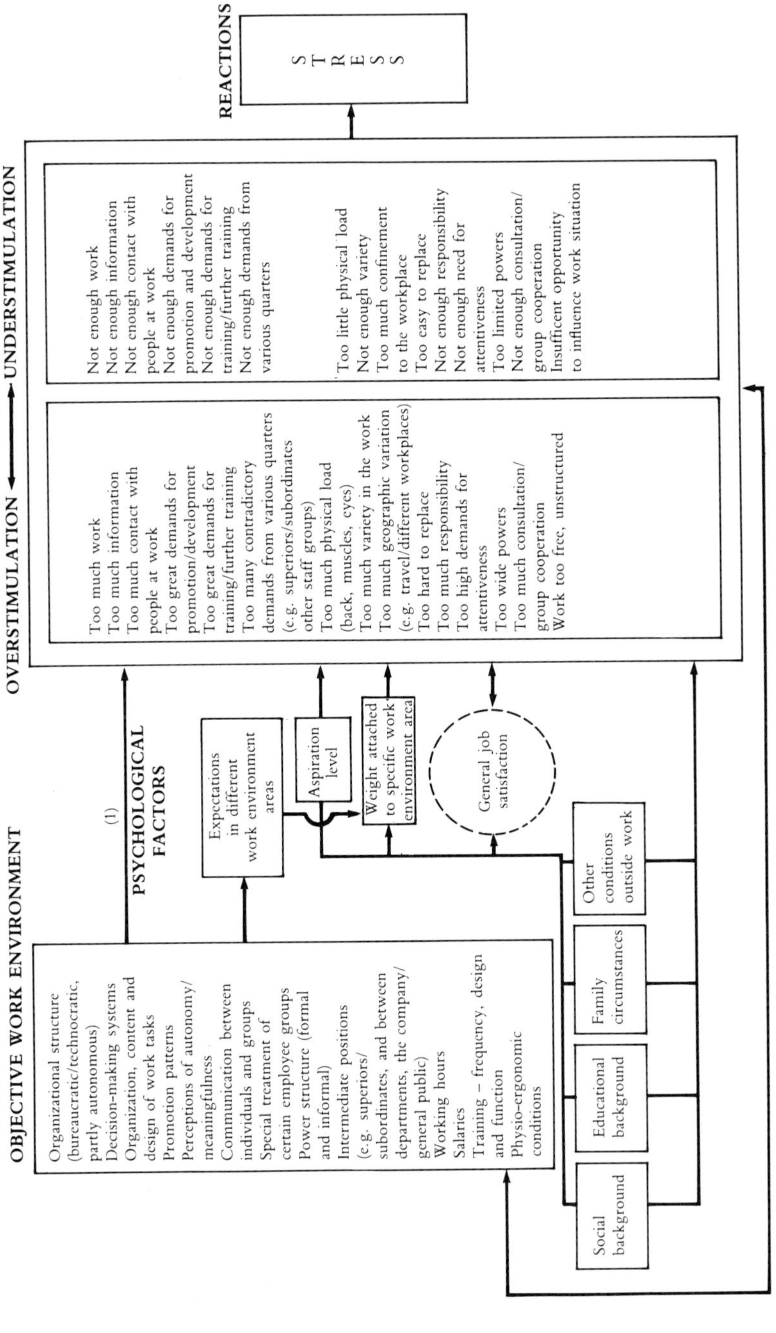

Figure 5. The relationship between computer technology, the work environment and stress (Bradley's model)

Key to symbols:
→ effect
↔ two-way effect

An important point of departure in the analysis of people's work and working conditions in a computerized society is the formation of theories that apply the concept of understimulation and overstimulation. The focus may be on psychophysiological and psychosocial aspects and on aspects of management theory, or combinations of all of these. This can be illustrated by reference to the theoretical models produced in the RAM Project and described in Chapter 3.

The areas at work, i.e., aspects of the work environment, that were examined in the RAM Project are in the boxed field at the left under the heading 'Objective Work Environment' in Figure 5. Each of the opposite pairs in Figure 5—the boxed field on the right—may be related to factors in the 'objective work environment'. Figure 5 reports only one way of reacting, namely stress. There are a number of other reactions that may be attributed to people's perceptions of work and their working conditions, which are in turn dependent on different objective factors at work. People frequently cite powerlessness and alienation, or their opposites, involvement and commitment. Other reactions include behaviour related to people's private lives and leisure activities. Both passive and active reactions can be identified (see Figure 2, Chapter 3). The models in Figures 1 and 2 in Chapter 3 form a whole; Figure 2 may be placed in the broader context that is depicted in Figure 1. As we are discussing stress and psychological strain, a brief comment on some of the terms that appear in Figure 1 is appropriate here.

The concepts in the model in Figure 1	Brief comments
Labour market and company goals	Efficiency
Society's norms and values	'The rational society'
Computerized information systems	Aids to automation
Technology	Mental and muscle power are replaced by machines, in pursuit of efficiency

It is not surprising that stress is a social phenomenon when we examine in this way the forces in play at the level of society and the organization. This may also explain why measures have more often been directed towards 'coping with stress'—using 'social support' to modify stress reactions, for instance—instead of attacking the causes of stress at different levels of analysis. Both these strategies must be applied. Here we shall follow a chronological sequence, beginning with the old, still extant, paper office and its implications both past and present in terms of stress, then focusing on the stress in traditional batch-processing environments, and finally examining stress in an online environment using display terminals.

2. Stress and the 'paper office'

Stress in the 'paper office': the picture needs no comment.
© Joint Industrial Safety Council

3. Stress and the batch-processing environment

The data processing system studied in the first company (see Chapter 4) was a batch-processing system. Some work tasks were done by people, while others were automated. In this process there is no direct contact with the computer. The work documentation often took the form of data lists instead of, for example, a text display on a VDT.

As mentioned earlier, it was mostly personnel from the Data Processing Department and operational departments (bookkeeping) who reported differences from other groups in terms of changes connected to the introduction of the EDP system. The staff in the bookkeeping departments had less education and lower qualification requirements, while Data Processing Department staff had more education. There were causes for dissatisfaction in both these areas of the organization, but these causes were different and produced different indirect effects, such as health problems.

Stress in a batch processing environment.

Photo: Tommy Olofson/MIRA

Some 2000 people, most of them women, worked in the operational departments, but the number has decreased, or is expected to decrease, at every stage of the process of automation. Here, staff reacted against the fact that the 'interesting' and 'stimulating' work tasks had been removed. Little variety remained in the work, and staff thought that they had been deprived of the responsibility they previously enjoyed. They felt insecure: 'the computer is taking over', they said, and were worried about being relocated. They were easily replaceable and this affected their self-esteem. Their influence had decreased, mainly the influence they had had on planning their work. Down time in the computer room often caused overtime and irregular working hours, making it difficult to arrange leisure and family time.

The pace of work, which was intensive and controlled by the machines, was seen to be too high; deadlines were an inherent part of the daily work and this served to increase workloads in all departments. What had formerly been a good level of cooperation had been worsened by work pressures. The introduction of EDP technology had aggravated certain physio-ergonomic problems, revealing substantial problems in this area in a traditional white-collar workplace.

The number of staff—mostly men—in the Data Processing Department increased. The staff in this department also felt more secure: computer-literate people were still in short supply. The Data Processing Department staff were not noticeably different from other departments in their attitude to job content. However, it was difficult for them to leave their job behind them when they left work and they often went to work even when they were unwell. They felt psychological pressure, partly due to the fact that there was no-one to stand in for them at work if they had to take leave. Their interchangeability was low, which had both positive and negative implications. The staff in the Data Processing Department had also lost some influence over their work—mainly over the pace of work and choice of work methods.

The computer system appeared to cause various types of psychological strain among the computer staff, one reason being the risk of errors occurring in the work, which made people nervous and worried about making mistakes. Errors had far-reaching consequences, affecting both the operational departments in the company and the company's customers. The Data Processing Department staff felt that they had to bear too much responsibility.

The Data Processing Department had become an isolated unit within the organization, and was described as being 'behind an iron curtain'. The Data Processing Department could be described as the organization's 'buffer department'. Demands, often contradictory, from various quarters appear to be concentrated or accentuated in this department.

Computerization resulted in work tasks/jobs that can be classified in one of four categories: 'too much', 'too difficult', 'too little', or 'too easy'. All of these categories caused stress, and the organization was divided up into sub-units with the attendant impact on personal relationships at work. From a crude organizational perspective, the operational departments had a number of work environment problems related to understimulation (both qualitative and quantitative), while the Data Processing Department showed signs of problems related to overstimulation (both qualitative and quantitative) (Bradley and Nilsson, 1976; Bradley, 1979).

In the process of reporting on the results, different measures were discussed with the local reference groups and with the central reference group for the project (see Chapter 4). Sub-reports from the project contained chapters on proposed measures at different levels aimed at 'preventive' and 'here and now' measures to eliminate stress problems.

It is now ten years since this project was carried out. New technology has been introduced since then and the old technology has been modified. Employee consultants have become involved. There has been intensive trade union activity and important government commissions have presented their final reports (see the last two sections of Chapter 4).

4. Stress and display terminals

Stress factors at work: Comparisons between display terminal users and non-terminal users.

In the investigations carried out at the insurance company as part of the RAM Project (see Chapter 5), the people interviewed were asked to give a direct description of a number of aspects in their work situation that broadly corresponded to the columns for 'understimulation' and 'overstimulation' in Figure 5. It is these work aspects that are referred to as stress factors, for instance too large or too small a volume of work, too much or too little contact with other people, and a work pace that is too fast or too slow. The estimates have been made on a scale of one to five, 1 and 5 being the poles on the scale and 3 being equivalent to neither too much nor too little. One breakdown of the material was a comparison between users and non-users of VDTs at work. This type of comparison is increasingly difficult to carry out: in certain sectors and companies most personnel already use a display terminal or will be doing so in the near future. Terminal users report more often than non-terminal users that their volume of work is too great, that they are given too little information and that there is too little consultation/group cooperation. Terminal users note more often than non-terminal users that promotion/

development prospects are too little, that they are too confined to their work station and too heavy demands are placed upon their physical abilities. None of the group's mean values for these factors is at the top or bottom of the scale.

Rather surprisingly, 'contact with other people' is also close to value 3 for terminal users, i.e., not too much and not too little. One explanation for this may be that many display terminal users at the company under study used the telephone a great deal. Also, contact among the staff often increases during the introduction of a new system. The factors related to security at work and the demand situation are also close to value 3. Non-terminal users consider more often than terminal users that they have too little responsibility and power.

Stress factors at work: Comparisons between display terminal users who work at the terminal for different lengths of time.

Comparisons were made within the group of display terminal users between people who use the terminal for less than thirty minutes a day and people who use the terminal more than two hours a day. In the following, the differences are marked when they are statistically reliable. The results in the previous section are repeated here to some extent: the relationship between the terminal user and the non-terminal user sometimes corresponds to the relationship between the people who spend a long time at the display terminal every day and people with shorter daily spells of work at the terminal.

People who spend longer at the display terminal every day report more frequently than people who spend a shorter daily time at the terminal every day that they are given too little information ($p<0.01$), that they have too few promotion and personal development opportunities in their work, and the educational/further training requirements are too modest. The people who spend more time daily at the terminal more often report a considerable degree of confinement to the work station ($p<0.05$), that they have too little contact with other people at work, and that physical loadings (e.g., on the back, muscles and eyes) are too great ($p<0.05$). More people who use the display terminal for longer periods every day feel more secure in their work. They think more often than the people who work shorter daily spells at the display terminal that it is far too difficult to find someone to stand in for them at work ($p<0.05$), that there is too little consultation at the workplace (see above), and that the management is rather too authoritarian. Terminal users at the company that was studied are predominantly women. It is also the female personnel who work at the display terminal for long periods every day.

5. VDT work, work loads and psychological strain

The concept of stress can also be examined indirectly by putting questions on other aspects of work pace, work load, psychological strain and various health problems. Here, the respondents were asked to what extent or with what frequency they suffer from certain complaints, or the extent to which they experience certain phenomena in their work. These areas have been analyzed with the aid of some of the indices set out in Figure 4, Chapter 5. Some of the more important stress-related results from the insurance company are summarized here. These results are discussed in Chapter 9 in relation to action taken.

Our comparison between terminal users and non-terminal users indicates that display terminal users experience 'stress' to a greater extent (age and sex as constants). The differences have been demonstrated in an index with the following content, expressed in separate questions:

- experience more strain in the job;
- think the work pace is too high;
- feel that the work is hectic and stressed;
- report undermanning;
- report too high efficiency requirements.

Furthermore, the display terminal users report more often than non-terminal users that:

- the work pace is too irregular;
- the work pace is dictated too much by other people;
- the work pace is too high at certain times;
- the demand situation has increased at work (the buffer role);
- attentiveness, accuracy and responsibility requirements have increased.

The display terminal is an instrument that helps us retrieve, process, and store information, and as such is a means of increasing efficiency. Making efficient and appropriate use of people's resources should be of benefit both to the company and the employee. A rapid throughput of work per unit of time may initially be perceived as a 'personal success' at work, but soon becomes a 'normal level' of performance. Thus, there is a risk that the work pace at display terminals will also affect our time rhythms and that its impact will spread to other areas, for example, individual contact at work and contacts outside working hours.

Terminal users stated more often than non-terminal users that they had general bodily pain and headaches, both at work and in their leisure time. The group of younger terminal users refer to health problems more frequently. Younger terminal users say that the introduction of terminals has increased the risk of physical and psychological complaints, and mention that they occasionally suffer from eye problems.

The Trygg-Hansa study shows that display terminal work can be reconciled with satisfying the individual's need for job autonomy and an interesting job content. The introduction of terminals does not need to lead to job depletion and the fragmentation of work tasks, as it did in the Postal Giro survey where a batch-type system was studied.

Comparisons between groups that spend different times at the display terminal each day illustrate the importance of discussing ways of combining display terminal work with other duties and avoiding constant work at the display terminal. Staff that work short daily periods at the display terminal:

- are more satisfied with their opportunities to use their own knowledge in their work (men and older staff);
- feel to a lesser degree that the buffer role (demands from different quarters that are related to their professional roles) is accentuated by the introduction of display terminals;
- complain less of general pain during working hours.

Staff working longer daily periods at display terminals have expressed various kinds of stress reaction, and have, more frequently than personnel who work for short periods each day at the display terminal, described their work situation as having the following characteristics:

- too little information;
- too little scope for personal development in the work;
- too low educational/further training requirements;
- too confined to the work station;
- too little contact with other staff;
- too heavy physical demands;
- too little consultation at work;
- far too difficult to find people to stand in for them.

Staff who spend long periods each day at the display terminal were more positive in the following respects. They feel that the introduction of display terminals has meant less risk of error and anxiety about making mistakes; more chances of correcting errors; and that the computer system has reduced the risk of error.

The attitudes of display terminal users with different amounts of contact with the general public/customers in their work have been summarized and discussed in Chapter 5. Terminal users with more contact with the general public/customers are more aware of the buffer role and feel more often than people with less contact that the job involves stress and psychological strain. They consider that:

- the work pace is too high, the work is too stressed, and there is undermanning;
- the work pace is too irregular or influenced too much by other people;
- the work requires far too much attentiveness and accuracy.

More than other departments, the Customer Service Department has a great deal of contact with the public/customers. The results indicate that the ancillary equipment to the terminal system has not enhanced the quality of personal contact in the job. On the contrary, too heavy a workload and stress appear to have reduced this aspect of the work environment. At the same time, an important goal of the terminal system is to facilitate contact with customers.

Different aspects of work pace (too hectic, too irregular, etc.) have been dealt with earlier. Chapter 9 examines in more detail the work environment phenomenon referred to as the intermediate position/buffer role, and indicates possible ways of dealing with, avoiding, or changing the buffer role.

Comparisons have been made between different professional groups, namely between secretaries and other female personnel (administrators) at the insurance company. The secretaries were more dissatisfied with those areas of their work that are normally seen as being crucial to job satisfaction. There appeared to be confirmation of this in the fact that their health was significantly worse—they suffered more frequently from stomach ailments and nervous complaints.

What opinions did the secretaries express on the terminal system and the data processing system in general? The secretaries reported less often than other female personnel that the introduction of the terminal system worsened their work situation. They had a significantly more positive attitude to the new technology in relation to effects of the terminal system on the pace of work, attentiveness and responsibility and effects on physical and psychological complaints. Thus they appeared to see the new technology as a chance of being 'released' from an unsuitable professional role and job design.

The amount of work at a display terminal, for longer or shorter periods each day, and the length of continuous viewing of a VDT, are dictated by the general content of the company's operation, people's positions in the organization and the structure and content of their jobs. In some departments in the company efforts have been made to make gradual improvements in job content by making changes in work organization. One example was in the Personal Insurance sector where five new insurance groups were designated and efforts were made to integrate the issuing of insurance policies with preliminary examinations, risk assessment, etc. The aim was to avoid specialized jobs and instead create positions that allow one person to deal with an insurance case from beginning to end, i.e., a type of 'job enlargement'. (See Chapter 9, section 4 for a description of the action taken at the company to prevent and avoid constant work at display terminals. These matters are also dealt with in the follow-up interviews in the last section of Chapter 5).

6. What did we lose and what did we gain when display terminals were introduced?

The people interviewed at the insurance company were asked to name the conditions that had improved as a result of the introduction of display terminals and three aspects that had worsened. The 'free responses' were subsequently grouped under different headings.

What did we gain?

Ninety-four people (38 per cent) of the total group answered the question about factors that had improved. Together, they reported a total of 175 work environmental factors that had improved:

more rapid (time-saving) work and less paper work	83%
simpler or easier, i.e., all information was supplied	32%
more reliable information, including updated details	22%
less paperwork, other aspects	16%
better customer service (customers need not wait)	15%

In summary, the dominant response was that display terminals speeded everything up—as they were time-saving.

What is worse?

Thirty-eight people (15 per cent) of the total group answered the question on factors that had deteriorated, and these people reported a total of fifty-seven work environment factors that had worsened. The list below is based on 'free responses' that were subsequently coded in groups. It is difficult to make interdepartmental comparisons because so few departments reported aspects that had worsened. The following factors were mentioned:

far too dependent upon the display terminal	34%
eye complaints, general	24%
sitting still, i.e.., tiring to sit for long periods at the display terminal, back fatigue, poor working position	24%
different kinds of 'stress', e.g., lack of time, work input can be checked	18%
queues forming at the display terminal or the terminal is too far away	16%
irritation caused by systems breakdowns	11%
type of work task (e.g., simple, monotonous)	11%
volume of work	11%

In summary, roughly half of the responses were related to stress!

Research at another Swedish insurance company also used physiological measurements to examine display terminal work (Johansson and Aronsson, 1984). Blood pressure, cardiac activity, and the secretion of adrenalin and noradrenalin were measured. It was noted that delayed response times and systems breakdowns caused stress and mental strain. Temporary systems breakdowns caused an increase in both blood pressure and the production of adrenalin.

7. *Conclusions*

The introduction of computer technology appears to stimulate the creation of professional roles that are characterized by either understimulation or overstimulation. A computer system should help create work tasks and working situations in which as few people as possible are overstimulated or understimulated. This must be an individual—subjective—assessment. Ideas have been put forward on a 'society of moderation', and perhaps we should start to think more about moderation at our workplaces and in our homes; about environments that avoid extremes, that have neither too much nor too little of the components that make up the environment and that nullify the extremes of the opposing pairs mentioned in the previous chapter. There are both quantitative and qualitative aspects to overstimulation and understimulation. One quantitative aspect is the amount of work to be completed per unit of time, while the complexity of the work, for example, is a qualitative aspect.

There is a risk that the future labour market will have one group of highly qualified, highly educated people who are difficult to replace, and a second group with very low qualifications and a high degree of interchangeability. The gulf between these groups will tend to widen. A third group, consisting of people not gainfully employed or receiving only unearned income (outcasts, marginalized groups or fringe groups) will increase, and the rate of this increase will depend on how we conduct the inevitable overall long-term restructuring of our society as a computer society (see the research findings from companies 1 and 2).

The following are reasons why work tasks and work environments that may expose people to one of the two poles of overstimulation and understimulation should be avoided in our society:

- the risk of stress (the individual level);
- the risk of a fragmented labour force (group/organization levels);
- the risk of marginalization and exclusion from the mainstream of society (individual, group and societal levels).

Demands and requirements on work environments and working life that arise from computerization can be formulated in different ways when seen from different viewpoints. Stress-related criteria have been discussed in this chapter. A list of general requirements for the psychosocial work environment from the RAM Project which should be considered when introducing computer technology, in particular so-called terminalization, is to be found in Chapter 9.

Some types of stress that occur in display terminal work can be counteracted with technical aids—aids that affect response times, availability, screen layout, standardization, etc. However, most stress problems are to do with work organization. Chapter 9 contains a more general treatment of different ways of approaching these issues. A special section in Chapter 9 deals with action in the company, and relates measures aimed at avoiding and preventing stress and ill-health when new technology is introduced more directly to the findings.

When introducing computer technology, not only the lateral division of a work task but also the vertical division of work must be considered when designing/changing work tasks and organizing work to suit people's needs and aptitudes. In structuring professional roles when new technology is introduced, consideration should be given to the five factors affecting job content listed in Chapter 1. Attention should therefore be given to some questions of principle:

- What procedures apply in related professions at the same level in the organization?
- What constitutes 'too much' or 'too little' at other levels?
- Which tasks should be computerized and which tasks are better done by people?
- Are there any work tasks that should not be computerized at all?

One must also attempt to define and make a concrete description of the desirable development of one's own professional role. Each of us will contribute our personal viewpoints, and we can then join forces to put the pieces together to form a whole. But all this requires knowledge, and the courage and self-confidence to apply it.

8. *The impact of the psychosocial work environment on leisure, family and health*

1. Introduction

A separate RAM report, no. 10, formulates some hypotheses on the relationship between the perceived work environment and the impact of work on leisure time and family life, and health complaints at work. The theoretical point of departure was my model of the psychosocial work environment (Figure 2, Chapter 3); this theoretical model has been translated into empirical terms that are set out in Figure 4, Chapter 5. The analyses are based on the statistical material obtained in the second field study of the RAM Project, presented in Chapter 5.

The groups of variables in the empirical model in Figure 4, to be found in the block headed 'Perceived Work Environment Conditions', will be the main independent variables. Several of the background variables listed in the blocks headed 'Work-related Background Factors' and 'Individual-related Background Variables' are also included in the analyses. The dependent variables are in the block to the right of the empirical model headed 'Health Problems at Work' and 'Effects of Work on Leisure Time/Family'.

The aim was to get an idea of the relative importance of the factors in the perceived work environment in explaining the effects of work on family life and leisure activities, and the physical and psychological problems people experienced during working hours.

The statistical method used to analyze the relationship between a dependent variable and a set of independent variables (also called predictors or regressors) was multiple regression analysis or—to be more exact—progressive multiple regression analysis.

The following is a short description of the variables in question and their operational definitions, and the results of the regression analyses. A more detailed description of these analyses is given in RAM report no. 10 (Bjerin and Martens, 1981).

In the model the concept of work environment is divided into the objective and subjective working environments (see Figures 1 and 2, Chapter 3). The objective work environment includes the company's organization, the decision-making system, the rewards and promotion system, the intermediate position/buffer role (role ambiguity), working hours and physio-ergonomic conditions. The subjective or perceived work environment includes, for example, attitudes to co-determination and decision-making, perceptions of self-determination, meaningfulness, opportunities for personal development and advancement in the work and the buffer role/role ambiguity.

There is interaction between the objective and subjective work environments, as may be seen in the model (primarily in the theoretical version in Figure 2, Chapter 3). This interaction is mainly through different psychological factors such as expectations and weighting. The concept of the psychosocial work environment focuses on the interaction between the environment and the individual (Bradley, 1979).

In the RAM Project, we attempted to develop the most reliable measurements possible for different aspects of the psychosocial work environment (see Chapters 4 and 5). The indices in Figure 4, Chapter 5 were therefore suitable for further analyses.

The basic model for the regression analyses is given in RAM report no. 10 (see Figure 6).

Figure 6 is a simplified version of Bradley's model, and constitutes the basic model for the regression analyses (see Figure 3, Chapter 4). It can be seen from the model in Figure 6 that it is not only the subjective work environment factors that affect the dependent variables; they are also affected by various background data, i.e., the individual and work-related background variables. Moreover, the background data have a direct effect upon the perceived work environment and on health problems. In a more detailed analysis of the effects of the subjective work environment factors on the dependent variables, the effects of the background data must be considered. This may be done in the regression analyses. The model in Figure 6 shows that the background variables and the subjective work environment variables are the independent variables of the analysis.

Independent variables

There are two types of background variable in the analyses: individual-related variables and work-related variables.

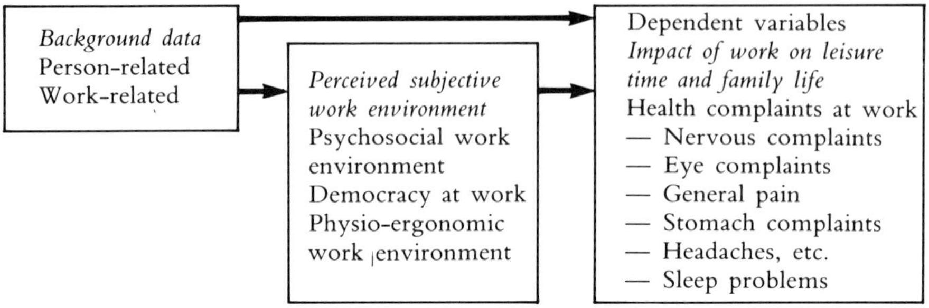

Figure 6. Model for regression analyses

a) Individual-related background variables:
  - Sex (two categories)
  - Age (six categories)
  - Civil status (three categories)
  - Basic education (four categories)
  - Children in the home, if any (two categories)

b) Work-related background variables:
  - Supervisory position (three categories)
  - Amount of time employed in the industry (six categories)
  - Contact with the public (four categories)
  - Working together with other people (three categories)
  - Working hours (three categories)
  - Income (five categories)
  - Contact with display terminals, number of hours/days (seven categories)
  - Perceived degree of dependence from the data processing system (four categories).

The work environment variables follow the division into 'blocks' set out in Chapter 5:

- Psychosocial work environment (16 indices)
- Democracy at the workplace (2 indices)
- Physio-ergonomic work environment (4 indices)

Dependent variables

The following dependent variables are used in the analyses:

- The impact of work on leisure time/family life (AF)
- Nervous complaints (HAN)
- General pain (HAV)
- Eye complaints (HAÖ)
- Stomach complaints (HAM)
- Headaches, etc. (HAH)
- Sleeping problems (HFS)

These variables are discussed in greater detail in the comments on the findings. A detailed description of the structure of the indices, more detailed questions, and a discussion of reliability, etc. are to be found in Bradley (1981).

2. *Findings*

The aim was to establish which working environment factors may explain different problems. Table 8.1 presents the analyses where all the work environment variables in the model are included and where the background factors are kept constant. A corrected R^2, that is, the part of the variance attributable to each respective group, is presented in each analysis for each group in order to get an overview of the explanatory values of the background factors and the four groups or blocks of predictors, namely the psychosocial work environment indices (S-indices), the democracy at work indices (DEM-indices), the physio-ergonomic work environment indices (F-indices) and the background variables.

What is the impact of the psychosocial work environment, democracy and ergonomic factors on leisure time/family life?

The indices for the impact on leisure time (AF) are based on the following questions:

- Do you think that your present job makes too heavy inroads into your home and family life?
- Do you think that your present position makes too heavy inroads into your leisure time activities?
- Do you think that you have too much overtime work?
- Do irregular working hours make it difficult for you to plan your leisure time?

(The scale of points was from one to five, high points signifying that the person interviewed does not think that work affects leisure and family life.)

Table 8.1 Beta-weightings for the work environment variables and multiple R^2 where both the background variables and the work environment variables were included in the regression analyses of the dependent variables

Independent variables (regressors)	Dependent variables	The effects of work on leisure/ family (AF)	Nervous complaints (HAN)	Eye complaints (HAÖ)	General pain (HAV)	Stomach complaints (HAM)	Head-aches (HAH)	Sleep (HFS)
Psychosocial work environment variables								
Interesting work – meaningful	SIME	−17	03	−15	−10	–	−07	05
Interesting work – human resources	SIMR	–	02	17	13	09	−04	15
Job autonomy	SSJA	08	08	05	−10	14	−08	−03
Contact opportunities at work	SKONT	−05	04	−10	09	−04	03	07
Cooperation with co-workers	SSAM	07	03	02	−07	−15	10	–
Salaries	SLÖN	06	−09	01	−05	02	−14	−05
Promotion – positive	SBEFP	–	–	05	−05	09	−09	03
Promotion – negative	SBEFN	12	28	19	01	21	02	11
Promotion prospects	SBEFM	−05	−01	03	−04	−02	11	−04
Information	SINFO	−11	03	−05	19	01	06	08
Intermediate position/buffer role	SBUFF	13	−07	−03	−04	–	08	−08
Work pace/work load	SARBT	36	40	31	24	29	21	23
Irregular work pace/work load	SOJA	11	−02	−03	02	01	03	−04
Job status	SAANS	05	18	−04	−02	07	04	09
Attentiveness/responsibility	SUPPA	01	−01	–	–	−22	−02	07
Efficiency	SEFF	09	05	13	06	–	03	01

Table 8.1 continued

Independent variables (regressors)	Dependent variables	The effects of work on leisure/ family (AF)	Nervous complaints (HAN)	Eye complaints (HAÖ)	General pain (HAV)	Stomach complaints (HAM)	Head-aches (HAH)	Sleep (HFS)
Democracy variables								
Company democracy	DEMF							
Company democracy – preferences	DEMFÖ	−04	−03	−04	−08	−03	02	−03
Work/leisure time democracy	DEMAL	12	04	04	–	12	08	−09
Work/leisure time democracy – wishes	DEMALÖ							
Physio–ergonomic work environment variables								
Work premises	FA	−02	11	06	09	02	11	05
Temperature/air	FT	−18	−04	04	02	−09	09	06
Ergonomic factors	FE	04	−01	01	07	04	−06	–
Odours	FL	10	06	17	04	04	03	01
100 . Multiple R^2		44	44	33	44	23	24	23
100 . Multiple R^2_{corr}		34	36	20	33	09	09	09

The psychosocial work environment factors best explain the impact of work on leisure time and family life (this group of variables explains 25 per cent of the variance). The significance of people's perception of democracy at the workplace and physio-ergonomic factors is negligible. It is mainly work pace and work load which affect the degree that people feel that work encroaches upon leisure time and family life. Whether or not people have an interesting and meaningful job also has some effect (see Table 8.1). Income is clearly the strongest of the background variables that help explain the effect of work on family and leisure time (see Table 8.2).

The following points summarize the findings:
Work was seen as encroaching on leisure time and family life by those who think that the work pace is fast and the work load heavy; those who think that the work is interesting and meaningful; and those with a high income. In an earlier regression analysis of a similar population, we found that work pace and work load have the strongest effects on the impact of work on leisure time and family life (RAM report no. 6, Bradley, 1979).

Among the salaried employees covered by the investigation, the faster the pace of work and the greater the involvement in the work and the higher paid the work, the greater the perception of a conflict between professional and family roles. The findings indicate that overstimulation at work causes fatigue after work, thus affecting the way people associate with their families.

Different studies have shown that heavy physical and/or repetitive work causes so much fatigue that by the end of the working day people are too tired to have anything more than superficial contact with their families (Gardell, 1979; Liljeström, 1973, Ch. 7). Both physical and psychological stimulation and psychological understimulation at work thus has a negative effect on people's opportunities to feel satisfied with their leisure time and family lives.

On the basis of the theoretical models and indices used in the RAM Project, Martens has made a further analysis of the relationship between work and leisure time/family life (RAM report no. 12).

What is the impact of the psychosocial working environment, democracy and ergonomic factors on health problems at work?

Here, health problems are based on subjective reports, and cover health problems on the job. For practical reasons, no other technique can be applied in this type of investigation. In the sphere of medical care, subjective reports of health problems are also important to further diagnosis. In stress research there is a high degree of correlation between perceived stress and physiologically measured stress.

Table 8.2 Beta weightings for the background variables

	The effects of work on leisure/ family (AF)	Nervous complaints (HAN)	Eye complaints (HAÖ)	General pain (HAV)	Stomach complaints (HAM)	Headaches (HAH)	Sleep (HFS)
Age	.10	.09	.00	−.20	.06	.08	−.12
Basic education	−.09	.03	.12	.15	.10	.01	.01
Time in the industry	.02	−.04	.04	.06	−.16	.09	−.13
Contact with the general public	−.04	−.00	−.13	−.19	−.09	−.16	.07
Work with other people	−.08	−.08	−.10	.14	−.02	.03	−.03
Working hours	−.01	−.06	−.07	−.10	−.04	−.00	.05
Income	−.28	−.10	−.13	.12	−.12	−.02	.00
Contact with VDT, number of hours/day	.15	.02	−.03	−.20	.01	−.12	−.02
Perceived dependence on computer system	−.01	−.04	−.07	−.12	−.15	−.01	−.09

Nervous complaints are the most 'sensitive' dependent variable for health problems (work environment factors explain 36 per cent of the variance). The next most 'sensitive' health variable is 'general pain' (the work environment factors explain 22 per cent of the variance.) The background variables have some effect on general pain. The predictors contribute 11 per cent to the explained variance. The predictors in the model have only a slight effect on other health variables.

Nervous complaints

The nervous complaints index (HAN) contains the following question:

> Have you been troubled over the past year by any of the symptoms listed below:
> - depression
> - nervousness, worry, anxiety
> - restlessness
> - irritability
> - periodic difficulties in getting anything done
> - fatigue
> - heart palpitations?

(The scale is from one to five points, 5 meaning that the person interviewed has never been troubled by these ailments and 1 that he/she has often suffered from them.) The psychosocial work environment factors are the strongest factors in explaining nervous complaints. Democracy at the workplace and physio-ergonomic factors are relatively unimportant. Work pace and work load is by far the strongest explanatory factor, but also Promotion—negative and Job status were fairly significant. The design of the work premises also has some effect, this being most apparent when only the physio-ergonomic work environment factors are included in the model.

In summary, there is a tendency to experience nervous complaints if:

- the work pace and work load are considered to be too high;
- people are critical of the promotion system, e.g., think that there is too much 'intrigue and gossip' associated with promotions, and that promotion creates a competitive atmosphere;
- people think that the status of their work is not high in the company and among their friends;
- people do not feel that they have any influence on the scheduling of breaks, holidays and working hours;

- people are dissatisfied with the design of their workplace [most of the people interviewed work in open-plan offices].

Bolinder and Ohlström (1971, pp. 84–5) found that among the membership of the Swedish Trade Union Confederation (LO) affiliates, work environment factors more than other background factors explained the perception of psychological strain. This is verified by the results presented here. Gardell (1979) found that among members of the Swedish Confederation of Salaried Employees (TCO) affiliated unions, all work environment factors contributed to nervous complaints. 'Influence on the work situation', 'Conflicts with superiors', and 'Self-determination at work' were the most significant factors.

General pain

The general pain index (HAV) contains the following question:

> Have you been troubled over the past year by any of the symptoms listed below:
> - pain and/or fatigue in the shoulders and neck
> - problems in the hand and finger joints
> - swollen feet and/or legs
> - back complaints?

(The scale is from one to five, with five points signifying that the person interviewed never feels general pain.) The psychosocial work environment factors do most to explain general pain. (Only 11 per cent and 6 per cent respectively are explained by the democracy factors and the physio-ergonomic factors.) The most significant of the work environment variables are 'Work pace and work load' and 'Information' (see Table 8.1). The background factors have a relatively large explanatory effect on general body pain. Of these factors, Terminal use, Age, and Contact with the public have the greatest weighting (see Table 8.2).

This confirms that the breakdown of the material carried out in the comparison between average values was appropriate (see Chapter 5). The following points summarize this relationship:

> People tend to suffer from general pain if:
> - they think that the work pace and work load are too high;
> - they think they are given too little information on events in the company, changes and conditions that affect their own work, or events in other departments;
> - they work for long spells daily at the display terminal;
> - they are older;
> - they seldom or never have contact with the public.

A hectic work pace, display terminal use, and age are factors that have also been pinpointed by other investigations as being related to general pain. On the other hand, research has shed less light on the link between pain and a fast work pace and lack of information.

Other research that has emphasized the connection between pain and the work environment is reported in RAM report no. 10. Because the installation of all the display terminals had not been completed at the time of our investigation it is possible that the work positions were not always ergonomically correct, a factor which is, of course, more noticeable when people spend long periods working at the display terminals. The interviews gave the impression that it was fairly normal for people working at display terminals to have both internal and outside telephone calls. Standard telephone handsets were used almost without exception, and this could probably explain some of the pain reported by display terminal users. These observations may be compared to Westin's (1975) observations, i.e., special load problems among personnel using VDTs for data entry, customer service and receiving orders. She maintained that faulty work positions are related to load problems for the eyes, head, neck and shoulder region and that psychological overloading coincides with an increase in muscular tension.

The connection between contact with the public and general pain may possibly be attributable to the fact that display terminal users who seldom or never had any contact with the public are either people who work mainly on data entry or who use the telephone in their contacts with the public (see above).

Eye complaints

The eye complaints index (HAÖ) contains the following question:

> Have you experienced problems of the type listed below at work over the past year?
> - Do your eyes get tired?
> - Do your eyes smart or show any other kind of irritation?
> - Do you suffer from eye complaints?
> - Do you ever have pain in or around the eyes?

(The scale is from one to five points, with high points indicating that the person interviewed never suffers from these problems.) Eye complaints can be attributed equally to psychosocial work environment factors and physio-ergonomic factors. (The proportion of explained variance is, however, not particularly high at 10 per cent). The strongest psychosocial work environment factor is work pace and work load. Of the physio-ergonomic work environment factors, the

Odour factor (that includes co-workers smoking) is the strongest explanatory factor. None of the background factors is particularly strong.

The following points summarize the findings.

People tend to develop eye complaints if:

- they think that work pace and work load are too high;
- they are critical of promotion-related competitiveness;
- they do not think that use is made at work of their personal resources and that the work is too simple;
- they think that their work is very interesting and enjoyable;
- they are bothered by odours or their co-workers' smoking.

The connection with the third and fourth points above may be interpreted as certain unqualified work tasks bringing about eye strain and that some work tasks are so interesting that people do not 'spare themselves'.

Terminal use was not a decisive factor in these regression analyses. The lack of any link between eye complaints and terminal use in the investigation may be explained in part by the fact that the terminal users mostly worked on group terminals, that is, several people shared one display terminal unit and the work design avoided constant work at the display terminal.

Stomach complaints and sleeping problems

The psychosocial work environment factors contribute most to explaining the variables in terms of stomach complaints (HAM). The following points are a summary of the results of the regression analyses:

People tend to suffer from stomach complaints if:

- they think the work pace and work load are too high;
- they regard promotion as negative in the sense that it gives rise to intrigues and competition;
- they feel the work certainly does not require too much attentiveness and involves absolutely no responsibility.

The connection between a hectic work pace and stomach complaints coincides with the argument of Bronner and Levi (1967, pp. 18–19) that there is a link between psychological stress and digestion problems and stomach ulcers.

The psychosocial work environment factors contribute most towards the proportion of the variance sleeping problems (HFS) that is explained by the work environment factors. The following points summarize the results of the analysis.

People tend to suffer from sleeping problems if:

- they think that the work pace and work load are too high;
- they think that their own work is too simple and that there is not enough scope for them to apply their own knowledge and skills.

In their investigation of blue-collar union members, Bolinder and Ohlström (1971) mention sleeping problems as one of the dominant psychosomatic symptoms of stress.

Headaches

The following question was put:

Have you suffered from any of the following symptoms at work during the past year? (HAH):

- migraine
- headaches

(The scale is from one to five points, 5 indicating that the person interviewed has never suffered from headaches and 1 that headaches are a common complaint.) The work environment factors explain no more than 10 per cent of the total variance and thus do not predict the symptoms with any real certainty. The background variables have no effect on headaches. The strong explanatory variable for headaches is work pace and work load. There is a tendency for people to report headaches if they feel that the work pace and work load are too high.

The regression analyses do not indicate that the amount of display terminal use would have any significant effects on headaches, even if the connection supports this. However, in broadly the same group there was a significant difference between display terminal users and non-terminal users, with more terminal users reporting headaches (Chapter 5).

3. Conclusions

Of all the work environment factors, Work pace and work load has the strongest effect throughout the various aspects of health problems that were studied. There is a link between a fast work pace and:

- nervous complaints
- general pain

- eye complaints
- stomach complaints
- headaches
- sleeping problems.

Most of the above variables have been used as symptoms of stress at work in research into stress. In our analyses the different types of health problems have proved particularly sensitive to the pace of work.

In the analyses we made the assumption that stress at work triggers off health problems. However, in theory the reverse may well be the case, that is, people with health problems are more likely to feel stress at work. But the most probable interpretation is that there is an interaction between perceived stress and health problems. This view is also put forward in Bradley's theoretical model in Figure 2, Chapter 3 in the two-way arrows between 'Subjective Work Environment' and 'Reaction'. Various views of health problems stemming from an individual characteristic or from work environment factors are discussed in Bradley (1979).

The work environment variable called Promotion—negative has some explanatory effect on health, at least in three of the variables for health problems. A negative perception of promotion is related to nervous complaints, stomach complaints and eye problems. Our results indicate that the present rewards system at work in the form of promotion has indirect consequences expressed as various 'stress symptoms'. The negative aspects of promotion that have emerged as a separate factor in the analyses could thereby be validated against different health problem factors. The analyses of 'Health problems at the workplace' support the perception that there are certain inherent negative consequences for the individual in the present pattern of promotion. A brief discussion on the limitations of the method is to be found in RAM report no. 10.

9.
Actions on computers and the work environment

The main thrust of the RAM Project is the analysis of the psychosocial work environment and the way it changes when data processing technology is introduced. At the end of this chapter we shall return to the view of the work environment presented in Chapter 3. In this view, people are affected by factors at national, company/organization and individual levels, factors that the individual may also influence. These different levels may also feature in discussions of the action to be taken.

What do the state, the company, the trade union and the individual do to improve work and work content when computers are introduced, and what action can be taken? On the basis of action already in progress in the field of computerization and the work environment, measures that should be given special consideration in the future will be considered.

1. The national level

At the national level, laws and agreements are important instruments. Work environment design is treated in detail in the rules and regulations produced by the National Board of Occupational Safety and Health in collaboration with trade unions and employers' organizations among others. In addition, the state allocates funds for training and information and for research in the area of computerization/working life. In articles and debates, the mass media also generate greater awareness and more recognition of work environment problems. Family policy is also a part of constructive work environment action when data processing systems are introduced.

State committees and computer policy

At the end of the 1970s two government committees were set up to investigate the problem areas highlighted in the RAM Project (*Computer Technology, Working Life and Communication: The Delegation for Long-term Research*, Bradley, 1977). Under the Ministry of Labour, a committee examined 'the impact of data processing technology on employment and the work environment.' Under the Department of Industry, a committee examined 'the impact of data processing technology on changes in trade and commerce'. A rough review of some thirty government committees, etc. related to EDP was compiled by the Committee for EDP and Electronics. This review was jokingly dubbed the 'jungle map'. It reflected the ambition level of 'society', the ambitions of Parliament, the government, the political parties and the trade unions and employers' organizations in terms of the level of monitoring as well as controlling and assessing the use of computers in working life and in the community in general. Both of these committees have presented their final reports (SOU, 1984, pp. 20 and 51).

For several years now Sweden has had a Minister for Computer Technology, served by a delegation on data processing technology. An EDP policy programme, 'Coordinated EDP Policy' (Parliamentary Bill 1981/82, p. 123), was presented by the government of the time. The next government followed up this proposal in a number of ways and, starting in the summer of 1985, produced three documents on computer-related issues. 'EDP Policy' (1984/85, p. 220) contains an overall policy view of the use of data processing technology. The view was that one should select computer systems and programmes that will enhance employees' skills and responsibilities. The reasons for encouraging broad-based training include giving women a stronger position in the labour market and improving the chances of handicapped people finding work. Priority was given to educational issues, both basic education and further training. The government wants to invest in developing Swedish software and to promote its use in education.

With the aim of improving democracy and helping the process of renewal in the public sector, the general public should be given easier access to information, and the public should have better and more efficient service. Data processing power should be decentralized. The main points covered issues of integrity and classified information and the importance of data processing technology in terms of Swedish vulnerability in times of both peace and war. Proposals were also put forward on the general structure of computer-related issues—the structure of EDP policy itself. In dealing with research and technical change, the proposal referred to a separate programme for information technology (1984/85, p. 218). There is a separate legislative proposal on the use of electronic data processing in government administration (1984/85, p. 225).

Laws and agreements

Legislation is one way of improving the work environment at the national level. What laws and agreements affect action on computer-related work environment issues? The Work Environment Act, the Co-determination Act and the Acts on Security of Employment, Sex Discrimination and the legislation to promote equality of employment among the young and among disadvantaged groups apply to these issues as well as to other action taken to create better work environments. The Work Environment Act (Ch. 2, para. 1) says:

> The work environment shall be kept in a satisfactory state having regard to the nature of the work involved and the *social and technological progress occurring in the community at large* [italics in original]. Working conditions must be adapted to human physical and mental aptitudes. The aim must be for work to be arranged in such a way that the employee himself can influence his work situation.

Para. 19 of the Co-determination Act, concerning the employer's primary duty to provide information, and para. 11 of the Act, on the employer's primary duty to negotiate with the trade unions must be followed when computer technology is introduced at a workplace. The primary duty to negotiate covers a broad range of issues, including various applications of data technology and systems development problems. In a number of sectors the Co-determination Act has been supplemented by national and local union-employer agreements.

In the white-collar sphere, which is the main area covered in this book, two documents were published that emphasize the importance of computer-related issues for the members of the Confederation of Salaried Employees' (TCO) affiliated unions. In its report to the 1979 congress, 'Computers and Working Life', the TCO presents a detailed description of the application of the Co-determination Act and the Work Environment Act in systems development. The document also discusses the limits to the trade unions' ability to invoke labour legislation. SACO/SR (The Swedish Confederation of Professional Associations) produced a debate paper, 'SACO/SR and Computerization', for its 1982 congress. This paper was a compilation of articles from researchers and active trade unionists. Jonasson (1985) reviews the Work Environment Act and the Co-determination Act and how they apply to the introduction of data processing technology.

Trade unions and employers

At the national level in the TCO/LO and SACO/SR there are special work groups for computer-related questions. Their job is to prepare policy and advise on these issues. As trade union confederations or professional associations, all three national organizations have made public their standpoints on data processing technology. With regard to the white-collar sphere, the report, 'Computers and Working Life', published in 1979 by a TCO working group, contains guidelines for action. The report dealt with the following main areas:

- the computer industry and the computer market;
- computerization and employment;
- work organization and information;
- computerization and the work environment;
- co-determination in systems development;
- training and research.

The report summarized the policy position for each of these main areas. The TCO approved an action programme for 1983–85 which covered employment, employee influence in the direction and design of automation, and demands for legislation on the work environment. The action programme for 1986–88 put forward demands that are also to be found in parallel action programmes. The programme stated that in several areas technology affects the chances of attaining important trade union goals such as full employment, a wage policy based on the principle of solidarity, equality between the sexes and shorter working hours. Two contemporaneous documents from the TCO are 'Creativity in Working Life: Professional Knowledge, Work Organization and the New Technology' (1985), a document produced by the Computer Council, and 'The Correct Approach to Display Terminal Work' (1984), produced by the TCO Work Environment Council.

SACO/SR published a debate paper and submitted a computer policy action programme to their 1985 congress. The programme includes the SACO/SR basic position on employment, professional qualifications, the consequences on the work environment, confidentiality, vulnerability, integrity and the policy on research, training and a set of regulations for data processing technology.

The trade unions—at both local and national levels—initiate research and investigations into the work environment and in the area of work environment and computers. Guidelines for local union work are set out in national agreements (e.g., the work environment agreement, the rationalization agreement, the co-determination agreement, the technology agreement and the agreement on equality). These agreements supplement the laws and ordinances (see above). Areas now covered by co-determination agreements include state and local government employees, banking and insurance.

A separate development agreement (SAF/LO/PTK) deals with the process of change brought about by the introduction of new technology in industry.

A new meaning of the term 'efficiency' has been defined in an agreement (the co-determination agreement for state employees). A number of equally important factors must be considered. In addition to the economic use of resources and the standard of essential services, the government agencies must take account of their employees' need for job satisfaction, a good work environment, security of employment, co-determination and the opportunity for personal development in the job. Particular attention must be given to promoting equality between the sexes.

The following is a brief summary of the Swedish Confederation of Employers (SAF) action on computerization and the concomitant work environment issues. In March 1968 the SAF published its first edition of 'Administration Rationalization through Management', a package of materials aimed to promote the decentralization of administrative development and rationalization. The 'AR book' that is part of this package has subsequently been published in large numbers (*AR: A Path to Development* [in Swedish]).

Fairly intensive collaboration between the Swedish Employers Confederation and the white-collar private sector trade union began in the autumn of 1969. A Joint Rationalization Council was formed by the Swedish Employers' Confederation and the Negotiation Cartel for Salaried Employees in the Private Business Sector (PTK). The publication of 'User Development' in 1972 was a supplement to 'AR through Management'. In 1973 the SAF-PTK rationalization council published 'New Rationalization'. Briefly, its theory is that the people who are responsible for a business operation must also be responsible for rationalization and development. In 1978 the SAF published 'Decentralization and Computers' by Persson and Strandler.

Over the years the SAF Occupational Medicine Department has been involved with various work environment issues related to the use of computers (primarily VDTs), work that has been directly coordinated with the work of the National Board of Occupational Safety and Health.

The mass media

The mass media play a vital role in spreading knowledge, contributing to a factual debate on computerization, and allowing contradictory and complementary views to be aired on the structure of our future work and our private lives. In 1980 an intensive debate took place in Sweden on the national referendum on nuclear power. The debate on the computer society of the future, which was at its height at the beginning of 1980, came to a halt. However, at that time no issue was put forward that was sufficiently clear-cut to be subject of a referendum.

There was a change in government in Sweden in the autumn of 1982, which heralded an increase in investment in Swedish technology. The debate focused on different ways of introducing a computerized information society and less on the type of society that it would become. The information put forward at the national level (see above) contained details that enabled the mass media to give extensive coverage to matters such as the long-term goals of an information society.

Today's debate on the impact of computerization embraces new groups, for example, people with higher education, partly because the development and use of so-called expert systems has got under way, involving entirely new groups such as engineers/designers and doctors in the debate on computers. The development of systems that aim to rationalize the work of systems development itself has affected the terms and conditions of work of systems development engineers. There is, for example, no strategy for an action programme on the advanced use of computers (AI or expert systems), and new types of work-related problems of an ethical and legal nature which have an indirect effect on the work of systems development engineers, and the psychosocial aspects of their work, are beginning to appear.

The National Board of Occupational Safety and Health

In collaboration with other organizations, including the employers' national organizations and the trade unions, this body formulates directives, codes of practice and regulations. National Board of Occupational Safety and Health publications include 'Psychological and Social Aspects of the Work Environment' (AFS, 1980, p. 14), on aspects of particular importance to people's psychological situation. It is divided into the following areas, all of which have a bearing on computerization/the work environment: work organization, personal and professional development, social contacts, a greater awareness of health and safety, physical and chemical factors, automation and the scheduling of working hours. The ergonomic aspects of display terminal work are highlighted under the heading of automation. Particular efforts should be made to avoid automated production methods which create isolated or unskilled work tasks or lead to a greater degree of confinement to the work station.

Two other codes from the National Board of Occupational Safety and Health deal with computerization: the code on solitary work (1982, p. 3) and the code on working positions and work movements (1983, p. 6). In June 1985 the National Board of Occupational Safety and Health published new codes for VDT work (AFS, 1985, p. 12) that came into effect in 1986.

Thus the general standards for a good working environment are embodied in the National Board of Occupational Safety and Health codes of practice (AFS, 1980, p. 14). Growing experience in the field of work environment/computerization gives rise to a need to discuss special amendments and formulations that may support the work in companies and government agencies to create a good work environment when computers are installed.

At the time this book is being written, medical research is investigating the links between work on VDTs and miscarriages. The burgeoning debate in the trade press has shown how important it is to conduct research into changes brought about by the introduction of new technology; research that integrates different disciplines and research methods (see also Chapter 11). Hypotheses on the correlation between miscarriages and VDT work that embrace more complex cause and effect chains, in which psychosocial factors and/or stress factors may be intervening variables, must be tested.

A rather cynical observation is that a suspected link between miscarriages and work on VDTs may at last break the principle of vertical and lateral division of labour that has been followed far too rigidly, and applied on the basis of limited economic models. In the computer era the negative impact of the structure of professional roles will become more apparent than ever before. This applies to conditions of work at both low and high levels in the hierarchy. New strategies are needed for the structuring of professional roles, strategies that embrace both low and high levels simultaneously—they are interdependent.

In its most recently published codes of practice, the National Board of Occupational Safety and Health states that the research carried out so far has not indicated any risk of injury from the radiation or magnetic fields generated by VDTs. In the spring of 1985 the Ministry of Labour contacted both employers and trade union organizations to discuss agreements to give pregnant women working on VDTs the right to transfer to other work, while waiting for more reliable research findings.

Education and research

Whether or not computerization will split the labour force into well-qualified and poorly-qualified parts or whether there will be an overall lowering or raising of qualification levels depends very much (at least in countries like Sweden) on both the role given in the community to training and on the application of new technology. Are we to use the educational system to adapt and modify the individual, or should more opportunities be provided for individuals to exert influence? The future function and content of education in a computer society must be examined more closely.

Educational problems and processes are a natural part of working life (Söderström, 1983). There may be questions concerning the planning, implementation and evaluation of educational measures. Examples include courses of different kinds, the introduction of new employees to the workplace and development measures such as job expansion and job rotation. Management development, management training and group development are other educational problems.

It is becoming more and more important to introduce systematic learning and a qualitative development of knowledge on the job for large groups of employees. Labour legislation also contains requirements for particular educational measures. Computerization has increased the need for educational research and development. Technical change creates the need for preventive educational action. This may take the form of further education aimed at improving general knowledge, providing specialized professional training, training for new work tasks or training as a step towards changing professional roles. To develop, spread, allot, and use knowledge in working life and private life will become increasingly important both as a practical measure and also as an area for research both in preparation for a computer society and in the computer society itself.

The report, *Work Environment and Display Terminals* (Bradley, 1981) puts forward some views on education to prepare for a computer society and education in a computer society. In summary, the following are recommended:

- recurrent training as a basic educational principle;
- special training programmes for certain groups;
- courses should be produced aimed at developing and improving basic professional skills, giving a better understanding of the company's operations and more knowledge of related occupations both within and outside the organization; these courses should reflect the basic objectives of the job;
- training as a means of developing and enriching the professional role;
- broad-based EDP training emphasizing both the societal and technological aspects of computerization;
- courses on the possibilities and risks of computerization aimed at people such as psychologists in company health care, and social workers in personnel administration.

A great deal has been done in the field of training in computer science and training for a career in data processing. The adult educational associations have started a series of courses in this field. Today, companies, trade unions and hardware and software suppliers offer a wide range of courses. The following are some of the main areas where training courses have been started to prepare the community for the advent of data processing technology.

An approximately eighty-hour course in computer science is given in the compulsory school system. Computer studies are to be integrated into other relevant subjects. At the upper secondary level computer studies are integrated into social sciences, mathematics and a number of other subjects studied in the different streams. The aim is to teach pupils that computer systems are always a part of a larger 'system' made by and for people and which can be changed. The current evaluation of this educational work is of strategic importance, both in establishing whether the goals can be met and the directives followed in this area and as a basis for determining the direction of future actions in this field.

A government programme of investment in broad-based education has been started. A trial project is under way on education aimed specifically at people with a low educational background. The project includes both a short introductory course in computer studies and a course in occupational training, specializing in the applications of new technology at work. The programme has reached a total of 8000 people, two thirds of the students being women.

The Swedish Parliament has decided to set up so-called renewal funds to introduce a large-scale and comprehensive training programme. Under this system, all companies with profits in excess of half a million Swedish kronor must deposit money in the funds, which will be used to train employees and to finance research and development work. Priority is to be given to groups of people who do not normally attend staff training courses in large numbers, for example, women. The public sector employers and unions have also agreed to introduce renewal funds to finance further education/development of the workforce and to provide better security of employment in periods of structural change.

Following the trend towards more advanced technology in different professional categories, the Labour Market Board has reallocated a part of its resources, earmarking funds for labour market training which places more emphasis on electronics and future technology.

So far, we have noted a strong upswing in employment in the computer industry, but this sector will become automated too. Some occupations—mainly at the lower levels of the computer industry—have already disappeared. In our society there is a great need of 'product development' in the service sector—services associated with traditional female knowledge and experience. We may need a parallel course of training for men in jobs in the community (whether they be paid or unpaid) which have hitherto been the province of women.

In the area of research, the faculties at universities and institutes of higher education, the Working Life Centre, the National Board of Occupational Safety and Health, the Research Institute at the Stockholm School of Economics, and the Industrial Development Institute have completed or are currently conducting research and development work on the organizational, psychological, sociological, economic and medical aspects of computerization.

Action on equality

The reforms and strategies in the area of family policy that serve to reduce the conflict between family life and working life are closely linked to constructive action on the work environment and computerization. This is also true of work for equality in the broad sense that is carried on at workplaces and in various groups and associations. Data processing technology affects women's work situations, partly because machines take over traditional female tasks and partly because new positions in the field tend to attract males. However, the traditional boy/girl separation of education and career choice has become less rigid in recent years. Computer studies now feature more prominently in the final years of compulsory education, and curricula are now designed to make computer studies more attractive to girls.

Computerization not only affects employment and occupational structures. It also changes the work environment itself, causing among other things qualitative changes in job content and in the relationship between work and leisure time/private life. Computerization affects both the extrinsic and intrinsic aspects of equality (see Chapter 1). Thus it is important that technological change in the field of data processing be monitored by bodies that deal with equality issues at both national and corporate/local trade union levels and also at the level of individual women who go out to work and who work in the home. It may be that a high degree of awareness of current trends will change the habits of women as a group so that instead of being mere 'observers' of technological change, they begin to occupy positions where they can exercise control and where technology becomes a means of supporting the struggle of both women and men for equality at work and in private life. In a separate RAM report (no. 11), I have focused on computerization and female personnel (see Appendix 2 and the list of reference material), and have presented a ten-point strategy for increasing women's influence in the area of computerization and for changing their conditions of work and life.

In 1984 the government allocated funds to attract more women to industrial and technical occupations (the so-called ten million campaign). The campaign was part of a long-term strategy aimed at broadening the female labour market and took the form of a nationwide project on three fronts: action to influence girls' choice of education and occupation; computer training for women; recruitment and further training measures in the labour market. Computer technology was an issue in most of these projects. The experience gained from the computer training project was applied in the broad-based training programme (see above). Experience from the campaign as a whole showed that if women are to take part in computer training programmes, then the courses must be specially tailored for women. In many cases, the activities introduced during the campaign have been integrated into the day-to-day work of public agencies and other organizations.

Much of the current discussion on action in the field of data processing has centred on the struggle for power and influence over the process of computerization. Scientific reports and more popular books analyzing various strategies for systems development have been published. They have advanced ideas for replacing the one-sided planning and implementation of the automation process that has dominated up to now with a democratic process (Docherty *et al.*, 1977; Docherty and Loman, 1983; Ehn and Sandberg, 1979; Sandberg, 1984). A synchronous process involving the recognition, structuring, description, and analysis of work environment problems related to organizational and psychosocial matters, which should be given particular attention when introducing computer technology, is equally important. Which are the dominant work environment problems? What are the good and bad aspects? Which aspects do we want to improve? What goals should we set? If these questions form the basis for work in study circles and courses on the work environment and work organization, then this will be an important step in the process of computerization. *Working Roles and the Quality of Life* (Bradley, 1979) was a contribution from the RAM Project to this type of activity. Another example is the course material entitled 'The Meaning of Work: A Report on Clerical Occupations', and study circle material aimed at strengthening solidarity at workplaces and study material from the insurance employees' trade union. Awareness and recognition of work environment-related problems and their underlying causes are gradually spreading. Interest expressed in action programmes is meeting with a broader response. Anxiety has been expressed that if the debate on computerization is conducted strictly as a power issue, we will merely have a change of elite, or possibly create two contemporaneous elites. The issue of power is a crucial one, but power must be given a meaning reaching far down the ranks at our places of work. An understanding of the interplay of individuals, organizations, and society is just as important as a knowledge of programming and systems analysis.

2. The individual level: Its importance

If people are to have a real chance of influencing their work situation, i.e., bringing influence to bear both as individuals and collectively on trade unions and politicians, they will need knowledge of and insight into the relevant issues. For that reason it is important that people try to understand the underlying causes of both dissatisfaction and well-being at work and accept that they are personally responsible for finding information and learning more.

When data processing technology is introduced at our workplaces, it is important that we express our wishes as individuals for future development and help answer the questions that were presented briefly in Chapter 1:

- What conditions of life (leisure time, family life, social life outside what we now call work) do we wish to have in the future?
- What are the desirable characteristics of a good work environment?
- What is a desirable content of our future professional roles?

In the analysis of the work environment, we pointed out the interaction of different levels of the environment. At a personal level the individual is affected by the environment, but the individual also has an effect upon his/her environment. We have been given a number of instruments at national and corporate levels and the conditions have been created for broad-based action to improve the work environment and work to bring about corporate change. But it is up to us as individuals to give our work its content, both in terms of desirable goals and ways in which we can work towards attaining these goals. Thus this interplay is also an important factor when we take action in this field. Just as important as the existence of laws and agreements is what we ourselves attempt to achieve on the basis of these laws and agreements.

The level of the individual is important in terms of the working environment. All organizations are made up of individuals. Phenomena at the corporate level often reflect phenomena at the national level and become part of our 'selves'. It is important—not least in terms of the work environment—to improve 'ourselves', to develop our personalities in different ways. When we work to bring about change, for example, when our workplaces are computerized, it is important that we feel 'good', i.e., that the scope for neurotic defence mechanisms is minimized. The work of bringing about change may sometimes have to be preceded by courses which improve self-confidence, so that we dare to voice our opinions and apply our experience.

3. The psychosocial work environment: Requirements and preferences

There are different ways to put forward our preferences and requirements for the work environment when our workplaces are computerized. For example:

1. General requirements may be set for an acceptable work environment, along with group-based requirements for a particular industry, a particular trade union, etc.
2. Checklists may be used for suggestions on which work environment issues should be considered in the process of computerization.
3. Questionnaires or interviews may be used to evaluate planned computer systems or systems already in operation.

General requirements

General psychosocial work environment requirements may be set out and put forward in many ways. It is important that we create professional roles that are not characterized by either understimulation or overstimulation (see Figure 7; the opposing pairs in the figure have also been used in the RAM Project questionnaire).

Under 'Conclusions' in Chapter 7, three reasons were given for avoiding work tasks and work environments that can be labelled as either understimulating or overstimulating. When planning the installation of new technology, it is important to create professional roles that avoid the 'poles' of overstimulation and understimulation, a factor that is related to stress (see Chapter 7). The RAM Project also produced some proposals for more general demands on the psychosocial work environment as follows (Bradley, 1981):

- a work pace that can be adjusted to suit individual capabilities (workload per time unit);
- a reasonably even work pace, not too dependent on other people's work pace and/or equipment;
- the opportunity to influence the structure and planning of work and working hours, e.g., the scheduling of breaks and the length and scheduling of display terminal use;
- the opportunity for training/further training and personal development at work;
- to satisfy people's need for personal contact and communication

OVERSTIMULATION	←→	UNDERSTIMULATION
Too much work		Not enough work
Too much information	S	Not enough information
Too much contact with people at work		Not enough contact with people at work
Too great demands for promotion/development	T	Not enough demands for promotion and development
Too great demands for training/further training		Not enough demands for training/further training
Too many contradictory demands from various quarters (e.g., superiors/subordinates other staff groups)	R	Not enough demands from various quarters
Too much physical load (back, muscles, eyes)	E	Too little physical load
Too much variety in the work		Not enough variety
Too much geographic variation (e.g., travel/different workplaces)		Too much confinement to the workplace
Too hard to replace	S	Too easy to replace
Too much responsibility		Not enough responsibility
Too high demands for attentiveness		Not enough need for attentiveness
Too wide powers	S	Too limited powers
Too much consultation/group cooperation		Not enough consultation/group cooperation
Work too free, unstructured		Insufficient opportunity to influence work situation
The needs of the individual	←→	The physical and mental work environment

Figure 7. The psychosocial work environment and possibilities of overstimulation and understimulation

as part of their daily work, they should have the chance to confer with others at work;

- there must be an adequate flow of information, both vertically and laterally, and the information must reach the people for whom it is intended;
- people whose work unavoidably involves them in the intermediate position/buffer role must be given special support and the negative effects of this position must be counteracted or prevented;
- physical strain on the eyes, neck and back muscles must be avoided;
- jobs should be created that have neither a low nor high replaceability factor, i.e., too high a degree of vertical and lateral specialization in the organization must be avoided;
- there must be a more even allocation of work tasks between the sexes, so-called extrinsic equality;
- new professional roles and changes in jobs and positions make use of both the traditional male and female experience, interests, and values, i.e., so-called intrinsic equality.

Checklists

Checklists may be used as an aid to help in making a rapid assessment of which aspects of the work environment should be improved and developed at the planning stage, or when a computer system is already in operation. This type of checklist was produced for the Postal Giro Centre survey (see Chapter 4). The checklist was subsequently revised on the basis of experience gained in the Trygg-Hansa survey (Chapter 5). This checklist is reproduced in Appendix 3.

Questionnaires and interviews

The work environment can also be evaluated by questioning personnel and the key people in the organization. An example is the three questionnaires produced and tested in the studies carried out at the Postal Giro Centre (RAM reports 1–4).

In the Trygg-Hansa investigation the quantitative questionnaires (approximately 400 questions) were developed from a questionnaire that was designed for the batch-type system so that it was suitable for display terminal systems (Chapter 5). The questionnaires were published in *The Work Environment and Display Terminals* (Bradley, 1981), with a detailed description of their content and structure and guidelines for their use in Chapter 2. They may be used to evaluate the environments of both operational and planned computer systems.

The areas covered by the questions are detailed in Chapter 5, and Figure 4 in the same chapter gives an overview of the indices which are placed in a theoretical context in the figure. The questions that comprise each index are set out in Appendix 1. Chapter 3 contains a table of four fields describing the methods applied in the RAM Project. Carrying out all the descriptions and analyses that comprised the sub-studies in the RAM Project is, of course, far too time- and resource-consuming a process. One must choose the areas and strategies that suit the purpose.

In a future computer society, the requirements and preferences for our total work environment will also need to be discussed in terms of the structure and content of our leisure time. The structures and definitions that apply to our lives today will change. For example, the principles for what is to be classified as paid work need to be analyzed now by various groups representing different interests in the community, and political standpoints must be adopted on the basis of their findings.

4. Action at the corporate level

Introduction

Here we shall return to the findings of the RAM Project, using the Trygg-Hansa study as an example, but keeping the discussion as general as possible so that companies which intend to computerize their operations—or have already done so—will be able to apply these experiences.

Our comparisons between display terminal users and non-terminal users show that display terminal users feel more ‘stress’ than non-users, keeping the background variables constant. This is expressed in several ways. Differences have been shown in an index containing the following questions (see also below):

- feel more strain at work;
- feel that the work pace is too high;
- feel pressure and stress;
- report undermanning;
- report too great demands for efficiency.

The display terminal users also say that:

- the work pace is too irregular;
- the work pace is too dependent on other people;
- the work pace is periodically too hectic;

- the buffer role (role ambiguity) at work has become more acute;
- the need for attentiveness, accuracy, and responsibility has increased.

The display terminal is an instrument that facilitates the retrieval and processing of information and as such is a means of improving efficiency. In itself, this does not need to run counter to individual interests; it all depends on what is meant by the term 'efficiency'. Nevertheless, the display terminal as an aid involves real risks. A rapid throughput of work per time unit may initially be felt to be stimulating, but can soon become the 'standard' of performance. Thus there is a risk that *the work pace of the display terminal could affect our future pace of life*, and its effects could also spread to other areas of our lives, for example, our personal contacts at work and during our leisure time.

There are special reasons for analyzing the work environment of a display terminal work station from the physio-ergonomic viewpoint, because it has been shown that display terminal users more frequently than non-terminal users report general pain and headaches, both at work and in their leisure time. The younger group of display terminal users is most explicit about health problems. They feel that introducing computers increased the risk of physical and psychological problems and they sometimes report that they suffer from eye problems. Our interviews revealed that an ergonomic study of display terminal work that involves the simultaneous use of the telephone should be undertaken (shoulder and back problems).

There are at least two ways forward. Either the total daily hours spent at the display terminal can be limited, i.e., the work can be 'formalized', so that periods of intensive display terminal work are interspersed with work breaks and periods spent on other duties that do not involve display terminal work. Alternatively, when a display terminal system is introduced, suitable modifications to work and work organization should be discussed. If certain work tasks are to be transferred to the system, alternative work tasks to be included in the job may be discussed, as well as the further training needs and/or which vertical or lateral division of labour principles may need to be changed (see also Chapter 1, section 4).

Terminal users claim more often than non-users that the work pace is too fast, too irregular, and that there are periodic work peaks. The balance between the volume of work to be done and the manpower available is a crucial one from the work environment viewpoint. Thus it is important in planning to follow changes in production volumes over long periods of time in order to predict changes in the workload and manpower requirements.

Finally, it is important to point out that work at a display terminal can be combined with the individual's need of self-management at work and an interesting job content. The Trygg-Hansa study shows that terminalization does not necessarily have to lead to job depletion and the fragmentation of work tasks that occurred at the Postal Giro Centre, where a batch-type system was studied.

There are methodological difficulties in measuring long-term effects. Both cross-sectional and longitudinal studies have their weaknesses. Our comparisons between groups with different lengths of experience of display terminal work must, therefore, be interpreted with some caution.

People with little experience of display terminal work are consistently more positive in their attitude to different psychosocial factors in the work environment and more satisfied with the influence they have on working hours than are people with longer experience of display terminal work. Moreover, they more seldom suffer from general pain in their leisure time. Findings like these may provide the basis for hypotheses on the long-term effects of display terminal work on the psychosocial work environment. These findings may also indicate that different types of work task should be transferred to the display terminals at different times, or that efforts to create a good work environment have met with more success during the more recent phases of terminalization.

The comparisons between groups that spend different amounts of time daily working at a display terminal underline how important it is to discuss alternative ways of combining display terminal work with other duties and preventing continuous display terminal work. Personnel who spend shorter daily periods working at a display terminal:

- are more satisfied with their opportunities to apply their own knowledge (men and older personnel);
- feel to a lesser degree that the buffer role (demands from various quarters related to the professional role) has increased as a result of terminalization;
- complain less often about general pain during working hours.

Personnel who work at a display terminal for long periods every day have expressed various types of stress reaction. They have, more frequently than personnel who work for short periods each day at a display terminal, named the following as characteristic of their work:

- too little information;
- insufficient opportunity for personal development in the job;
- too low educational/further training requirements;
- too confined to the work station;

- too little contact with people at work;
- too high demands on their physical capacity;
- too little cooperation at the workplace;
- far too difficult to find replacements.

Employees who work for long daily spells at the display terminal were more positive in the following respects. They feel that the display terminal system means: less worry and anxiety about making mistakes, more opportunities to correct errors, and that the computer system has reduced the risk of errors occurring in the work.

One may ask whether people need to work for long periods each day in order to benefit from the obvious advantages of the display terminal system in terms of storing and processing large amounts of information with greater accuracy, thus serving to reduce the risks of work errors for the personnel. If this is the case, more internal training in the use of display terminals for people who use the terminals for a short time daily could be a remedy.

The attitude to their work of display terminal users with different amounts of contact with the general public/customers has been summarized and commented on in Chapter 5. Terminal users with more contact with the public/customers feel that the buffer role is more onerous, and feel more often than people with less contact with the public/customers that work involves stress and psychological strain. They feel that:

- the work pace is too hectic;
- there is too much stress in the job;
- they suffer from undermanning;
- the work pace is too irregular or too influenced by other people;
- the work requires far too much attentiveness and accuracy.

This applies primarily to the Customer Service Department, where staff have a great deal of contact with the public/customers. The results indicate that the ancillary equipment that a display terminal system requires does not promote personal contacts at work. On the contrary, too high a work load and too much stress seem to reduce this work environment factor. At the same time an important aim of installing a display terminal system should be to facilitate customer contact.

In addition to these work pace factors (too fast, too irregular, etc.) the work environment factor referred to as the intermediate position/buffer role is important. Employees with more contact with customers can easily find themselves in an intermediate position or buffer role (see Bradley, 1979, Ch. 6). A fast and irregular pace of work combined with the need for accuracy and attentiveness serve to accentuate the

The buffer role: contradictory demands.

Photo: Sven Oredson/Bildhuset

buffer role. Dissatisfaction with this state of affairs appears to be expressed in dissatisfaction with pay conditions and influence at work. Various ways of 'lightening' the buffer role, for example, by reviewing the manning level in these positions, both in general and over time, should be introduced.

Working Roles and the Quality of Life (Bradley, 1979) examines general ways of avoiding or modifying the buffer role, including:

- attempting to balance the business objective against the personnel and material resources available;
- attempting to minimize the contradictory content of the work objectives;
- checking that the staff's responsibilities and powers coincide;
- describing jobs, for example, when advertising positions setting out the conflicting demands that are an inherent part of some jobs;
- having an open discussion on whether or not staff have to deal with irreconcilable demands and conflicting expectations, and trying to take up these problems at the right level.

In the future it must be important, both from the work environment viewpoint and, more specifically, in terms of cooperation at the workplace, to establish the following:

- what other entities have holdings in your company (people, groups, organizations)?
- what are the objectives of these shareholders?
- which of these shareholders have the greatest effect on your own work and your own work situation?
- which of these are most important to the employees' superiors and subordinates?
- how do new laws and regulations affect these shareholders?
- what inherent conflicts are there?

In the preventive work carried out by company health care organizations, an evaluation of data processing technology should be broadbased and tap the different areas of expertise to be found in the company health service. The work of the company health service is both preventive and 'caring', the latter being of great interest when a data processing system is introduced.

In general, the frequency of the symptoms described in the investigation is low. However, there are differences between employee groups, for example, display terminal users allude more often to problems such as general pains and headaches at work and during leisure time than do the non-users. Of the work environment factors, 'Work pace' and 'Work load' have the strongest effect on all the health problems that were examined, including headaches. These results indicate a need for both preventive and corrective action here and now.

These findings are also of value in diagnosis in the special work done by the company health service. It is important in systems development work to integrate the objective of creating a good work environment in terms of influence, work pace, job content, the structure of the organization, personal contacts at work, information to personnel (see above).

The measures planned and the action taken at the company have been discussed in Chapter 5. In brief, action covered the following:

- the structure of the company's organization and the organization of positions in the company;
- the organization of systems development work;
- information (from both the company management and the trade union);
- training;
- the company health service;

- ergonomics;
- research and external consultation;
- trade union work.

Organizational structure, the structuring of work and professional roles (primarily to prevent continuous work at the terminal)

As the job structure at the Postal Giro Centre was discussed in Chapter 4, only the Trygg-Hansa organization will be discussed here. At the time of writing the research at Philips Electronikindustrier described in Chapter 6 was still in progress.

At first, terminal-based jobs were mostly pure data entry, but the system was extended to make terminal work an integrated part of the entire insurance operation, so that the staff now perform a variety of duties, alternating between work that involves customer contact, contact with field personnel, using tariff tables and handbooks and using the terminal as an aid in this work.

Some departments tried to improve work content by reorganizing and introducing new routines. One example is the Personal Insurance section, where new insurance groups have been developed. The job of writing insurance policies was combined with pre-checking, risk assessment, etc. The object was to avoid specialization and instead create jobs where the staff could handle an insurance policy from beginning to end, i.e., a type of 'job expansion'. At first, these changes only involved the policy writers, but at a later stage in this project the staff who worked on altering insurance policies asked to be included. By 1976 there were three departments within the Personal Insurance sector, one for each market area.

By and large, these changes were well-received, although one group of slightly older staff was worried about having to learn too many new routines. But nobody was forced to adopt the new routines. As a result of the reorganization, more people began to use terminals and the work was arranged so that nobody had to spend a long time at the terminal every day. Before the reorganization, these groups were far more vulnerable to the effects of work load peaks, when staff were on sick leave, etc. But the reorganization made it possible to 'loan out' staff to other units.

In 1980 another trial project was started in the Personal Insurance section. This concerned the management system. Five people in executive positions exchanged jobs with each other. This 'job rotation' allowed the managers to act as executives for each other's departments, which was expected to improve inter-departmental cooperation and reduce inter-departmental competition. Some changes had to be made in the management role and at the decision-making levels in the Personal Insurance section. Some comments on the experience gained from these projects were made in the follow-up interviews carried out in 1985 (see Chapter 5).

5. Roles

When data processing technology is introduced at a workplace, both the lateral and vertical division of work tasks must be considered in designing or changing job specifications and organizing work that will coincide with people's need and capabilities (see Chapter 1). When designing professional roles, one should consider:

- whether other work tasks in the production chain (at the same level) should be included;
- whether planning and control tasks (taken from a different level) can be included;
- whether some work tasks should be mechanized;
- which work tasks should be mechanized and which tasks should be done by people;
- if data processing and teletechnology should be used at all for some functions.

Empirical research shows that both overstimulation and understimulation are present (affecting both individuals and staff groups) as processes of alienation at the lower levels of the organization. Taken together with a critical analysis of management design and the executive role, this research supports the theory that these five strategies should be given greater emphasis in work organization and job design.

If properly designed, computerization can help solve a number of current problems in the management and executive function and in cooperation and the role of the co-worker. The three basic functions of today's executive—administrative, executive, and social—are often regarded as too heavy a work load. Companies are finding it difficult to recruit managers to some sales positions, one reason put forward being that the 'carrot' of a higher salary is no longer an incentive.

The people considered to be the best management material are reluctant to accept promotion to an executive position, partly because of the conflicting demands of the management role (the buffer role). Many people prefer to be 'specialists' on a lower salary scale. Alternative forms of management, such as taking a management job for a limited period, are under discussion. The company would like to see more of this 'mandate period' approach to job rotation at management level. Introducing changes in management and the management role is seen as an important way of preventing continuous work at terminals (see above).

The introduction of data processing technology has changed management styles and the executive role, and the company is moving towards an organization with fewer levels in its hierarchy. The company's aim was decentralization, and installing a terminal system is seen as the first step in this direction. Telecommunications technology may possibly be used for the same purpose in the future. A policy started early in 1981, to provide computer power locally, may increase the possibilities for a more decentralized operation, and this would have an indirect effect on work and job content (see also Chapter 5, follow-up interviews, 1985).

The organization of systems development

Following the 1976 reorganization, the philosophy of the insurance company was that systems development should be controlled by the end-users, and so-called Planning/AR units were set up for this purpose. Technicians and users sometimes have divergent directions, but they both aim to develop 'user-friendly' systems.

Systems development is carried out as a project, supervised by a group consisting of representative of the system users, the systems engineers and two trade union representatives. Work groups consisting of representatives of all users—mainly from the department in charge of the project but also from the Systems Analysis Department—are under the supervisory group. There is also a handbook describing routines for documentation, the time-schedules for crucial decisions, what material must be produced as a basis for decisions, etc. This form of project work became more firmly established in the company while we were working on our research project there.

Information (from the company and the trade union)

Because the Administration Development Group, the decision-making body for 'development issues' in the company, is a joint group, both the company and the trade union are able to provide information. The way the projects are organized (see above) gives the department in question the opportunity continually to monitor and influence the project through its representation in the supervisory group. Details on ongoing development projects are published in the staff journals.

Education and training/retraining

The scope, content, and structure of training courses are of strategic importance in a computer society. Some views on training in and in preparation for the computer society were expressed earlier in this chapter. The points discussed there can also be applied to companies and other organizations and to company-sponsored training. Here we shall consider the courses run and materials produced while we were doing research at the insurance company.

A handbook, *Display Terminal Work*, was published by the VDT Work Study Committee, which consisted of representatives from the Systems Analysis Department, Accounts, Health Care, and the Consultants and Architectural departments. The committee worked with representatives of all personnel categories that formed the target groups for this training programme. This material was aimed primarily at the planners: terminal work stations should be planned together with the users, the employer, and the safety representative. The issues dealt with in the handbook were:

- general company regulations;
- visual distance and eyeglasses—instructions;
- functions analysis;
- workplace design;
- lighting;
- a catalogue of furniture;
- the technical equipment;
- laws and agreements;
- practical information on the use of VDTs;
- the Customer Service Department as a workplace;
- the Claims Department as a workplace.

The handbook, which dealt mainly with the ergonomic aspects of terminal work, was used in other companies in the insurance industry and in other sectors.

The general company regulations deal mainly with the design of the company's terminal and fiche work stations. The following are some of the points covered in these regulations:

- To avoid eye fatigue and vision problems, arrange work so that personnel alternate work with rest breaks or duties with more conventional visual requirements. The need for this type of job rotation and rest breaks when working on a terminal depends on the design of work methods, the amount of reading from the screen that must be done, etc.
- Pure key-punching data entry should be avoided. If this type of work predominates, then the operator must stop work as soon as any problems arise or, after two consecutive hours of work

at the terminal, spend the same amount of time on other duties.

- To avoid ergonomic loads such as document-keyboard-screen eye movements for each transaction, a minimum of thirty hours training in basic typing skills is recommended for people who use terminals frequently in their work.
- If the user is to influence and rationalize his/her own work tasks, he/she must be able to assist in, or be a party to, the planning and design of his/her work routines.
- An acceptable work situation at a properly designed terminal work station requires that the original design and organizational proposals are checked and monitored.

The following principles were applied in other training courses and computerization procedures in the insurance company:

- All new employees completed a basic general EDP course;
- Planning and training courses were a part of every major development project;
- Industry-specific and management training courses included material aimed at giving employees a broader knowledge of EDP;
- Company management and the trade unions arranged seminars and discussion evenings on the problems of computerization.

Plans were made for basic courses for systems engineers and programmers on the following:

- The work content and work situation of insurance company employees;
- The best way to communicate with end-users.

Company health care

The activities listed above applied to company health care work at the time the RAM report, *Work Environment and Terminals* was written.

The insurance company's health service was involved in the following activities on computerization:

- medical check-ups;
- compulsory eye tests for terminal users;
- ergonomic checks of the workplace/workplace design;
- assistance in producing the handbook on terminals;
- training work supervisors and safety representatives;
- daily gymnastic sessions during work breaks on each floor of the office building;

- physiotherapy/back treatment arranged by the company physiotherapists for certain staff groups;
- outreach work and personal counselling on work positions, etc.;
- publishing information in the staff journal on correct work positions, etc.

In the follow-up interviews carried out at the company that are recorded in Chapter 5, we met a physiotherapist. Most of the activities she reported then are still going on, and the following supplementary physiotherapy programmes have been added in recent years:

- a neck class;
- a relaxation group—for people suffering from neck complaints, headaches, etc. in stress and tension situations;
- back treatment—for all patients who had been recommended physical exercise;
- voluntary group gymnastics sessions.

Ergonomic factors

In its Office Administration Department, the company worked out guidelines for an ergonomically correct workplace. Architects, physiotherapists, and users worked together to produce guidelines for designing so-called type-specific workplaces. Special attention was given to terminal work stations and a handbook was produced on this subject (see under 'Training').

Research and consultation

In addition to the RAM Project that was carried out at the Department of Sociology, University of Stockholm, the Trygg-Hansa insurance company also took part in a project on open-plan offices and in research on stress reactions to changes in working life, run by the Department of Psychology, University of Stockholm (Anita Rissler). External consultants were used for special tasks in the areas of systems development/AR, and training and management development among other areas. Several other companies were involved in work on administration, personal advancement and development in the job, and personnel planning.

Local trade union action

The insurance industry was one of the first areas in which computers were introduced and widely used, and the intensive work done on action related to computerization in this sector provided inspiration to other sectors that faced computerization. Some of the main events at the national union level are relevant to the description of trade union work at the insurance company.

The National Federation of Insurance Workers (FTF) developed an action programme for computer-related issues (*Co-determination in Advanced Technology*, National Federation of Insurance Workers, 1978). The aim of this programme was to support trade union representatives working in local project groups and groups involved in decision-making in the field of systems development. This programme included a description of the way that systems development work should be structured. The models are taken from the Swedish Institute for Systems Development guidelines for administrative development, and include a number of requirements to be considered in project work. A number of questions were also included which trade union representatives could use to assess the different consequences of introducing a particular system and help them work out alternative plans of action.

The action programme also contained a number of work environment standards for the introduction of terminals and requirements for the project's work, for example:

- the project should not last longer than 12–18 months, otherwise the overall picture of the project will become blurred;
- the users should be represented in the project groups from the start of the project;
- the users who join the project should be given training and information to prepare them for this work and they should be freed from their normal work duties;
- trade union representatives should sit on decision-making bodies in the group;
- the consequences for the personnel involved should be analyzed, for example, the effects on the number of jobs and on job content.

This action programme was circulated to the union membership for comment, and their response formed the basis of an official comment by the trade union's 1980 national convention, which issued the following directives to trade union officers:

- prepare a draft agreement on rationalization/technology based on the action programme;
- examine and decide upon advisable technological changes, etc.;
- demand, through the Swedish Confederation of Salaried Employees (TCO), a temporary halt to further computerization;
- decide as soon as possible how to press the demand to halt computerization, and decide on the level of computerization that should be maintained and how quickly computerization should be halted.

In the response from the membership, the greatest support was for the issue of the speed of technological change. Even though a temporary halt might adversely affect the present level of prosperity, trade union members insisted on a temporary halt in order to establish control over the process of computerization.

The FTF 1982 national convention adopted an action programme for the continuing trade union work on co-determination. Technical change and rationalization were dealt with as a sub-area in this programme. The convention also adopted a final action programme specifically for technology. The programme is a compilation of two previous programmes on *Co-determination and Advanced Technology* (1978), an action programme on computer-related issues, and *Supplement to the Action Programme* (1980).

Several local trade union committees appointed special 'computer technology reps', following the decision taken at the 1980 convention. Their task is to monitor the development of computer technology in the company on behalf of the local trade union committee. The role of this representative is set out in *Guidelines and Instructions for Computer Technology Representatives,* published by FTF.

The trade union's data programme for 1982 includes a three-point summary of policy:

- the employer's responsibility to negotiate on the introduction of, and changes to, the data processing system;
- the trade union's need for training to enable them to make a critical analysis of systems development proposals, etc.;
- the need to use expertise at national level for local support and representation.

A chapter in the action programme is devoted to project work and contains proposals for a ten-step process for dealing with computer-related issues in decision-making bodies for administrative rationalization. There is also a list of important requirements to be considered in project work and a number of key factors for analyzing the consequences of any action taken. The programme aims to provide guidance for trade union monitoring of computerization and its impact and to help make local trade union representatives better equipped to safeguard the interests of their co-workers when representing them in project groups and other decision-making bodies concerned with data processing.

The 1982 action programme notes that a stop on recruitment, relocating employees, and early retirement were measures used to 'resolve' the employment problems arising from computerization in companies. The report states that in future, problems caused by structural change should be resolved in a different way. It emphasizes that even if job content may have to be modified, the quality and level of knowledge at work must be maintained or increased despite the impact of computerization programmes. The report also calls for a stop to the elimination of jobs. The action programme puts a strong case for a vigorous product development programme that can create new jobs so that the industry will grow and employ more people.

After the trade union polled their membership on technological change, certain additions and modifications were put forward in 1979/80 on:

- a limit on time spent at terminals;
- 'no' to home-based terminals;
- a technology agreement;
- trade union control of the pace of change.

On the last point, the report states that demands at a national union level for trade union control of the pace of computerization remains, but the much-criticized phrase 'temporary halt' is no longer of interest. This phrase was removed by a decision of the 1982 convention. The FTF summarizes its advice to people responsible for computerization as follows:

> On the basis of our experience of events since 1977, we should like to particularly emphasize that in your role as local computer technology representative you:
> – keep up to date in general assessments of advances in computer technology in your discussions with management, etc.,
> – note each stage in the introduction of new technology and its effects, for example, problems with terminals and psychosocial factors, such as the effects on companionship between co-workers,
> – are alert to opportunities for increasing trade union control of technical change,
> – attempt to stimulate new ways of developing new products and creating jobs in order to counteract the future effects of rationalization in the company,
> – demand trade union representation and the facilities you need to apply these principles in your trade union work. (*Guidelines and Instructions for Computer Technology Representatives,* FTF)

The Work Environment Act is sometimes seen as a better instrument than the Co-determination Act. In 1983 the insurance industry's employers' association (FAO) and the FTF signed a work environment agreement for the insurance sector. Trygg-Hansa also signed a local agreement as a result of which courses are being run on the work environment.

The local FTF staff association involvement in computerization has been discussed under several of the above headings. The following are some brief additional comments:

- Since 1975 the association's committee has constantly monitored events in the field of EDP. The association has represented the personnel in project groups and in the executive group holding overall responsibility for administrative development at Group level. Personnel representatives were also included in a separate reference group to train system users that was established in the same year.
- In 1976 a special 'terminal environment group' was set up to monitor the impact of introducing terminals. Two staff representatives are included in this group.
- In 1978 the economy section of the staff association held a 'hearing' with the head of the Systems Analysis Department.
- A special group, including staff representatives, was appointed in 1977 for investment controls. In the same year the association arranged a 'theme evening' on computerization and the work environment.
- In 1979 the subject of computerization was included in the so-called 'union supermarket'.
- In some departments of the company, the trade unions meet a couple of times a year to discuss computer-related matters.

Local union workers report great difficulty in generating interest among employees in data processing issues: 'I have only ten years left' is a common response. In 1980 each department had a union representative involved in issues related to working conditions. However, some departments find it hard to get people to volunteer for this job. Some consultative groups used to exist, for example, at the time the company moved into its new offices. Management groups with personnel representation had been set up a couple of years before the Co-determination Act was passed. The type of work done by the trade unions and company management since 1980 is described in the follow-up interviews.

At the final discussion on action that we held in the spring of 1981 both the unions and the company put forward a number of ideas on the effects of the temporary halt in computerization in 1977 (see Chapter 5). Four years later, in 1981, there was a positive reaction to some of the effects of the events of 1977. It was a 'breathing space' in a company that was in a process of constant change, involving mergers, relocating to open-plan offices, etc. It allowed the company to apply and assimilate the knowledge of systems development and ideas on the structure and content of work that had emerged. To avoid depleting work tasks, jobs based solely on data entry or data retrieval duties were avoided. Efforts were made to influence the vertical division of labour in some parts of the organization (the management system). The forms for working in project groups on systems development were improved to give better contact with the end-users (see previous section). Finally, the hardware suppliers 'adapted' to current developments. They had had a very high level of ambition with regard to the design of certain specific systems which, partly due to the very high costs involved, subsequently proved impossible to carry out. The 'adjustment' meant using more 'ready-made' products. At the same time, in 1981, it was claimed that the experience gained from the high level of ambition prior to the halt in computerization had been of some benefit, both in the systems development work immediately following the halt and in the future process of computerization. For different perspectives on the dramatic sequence of events of 1977 see the interviews reported in Chapter 5 with the representatives of the Personnel and Data Processing Departments, the company health service and the trade union.

10.
Future research

1. Important areas for research

Computer technology and its use helps to determine the underlying structure of working life and society in general. But computer technology is also controlled by people. The pace of computer technology development is accelerating and it is spreading into more and more areas of society. Technical conditions are under constant change, and different combinations of telecommunications and computer technology provide new conditions for the structure of working life. Even though we have been given some means of controlling technical advances, we must continue to analyze changes in human conditions. This is true not least of social sciences and humanistic analyses aimed at establishing, for example, whether the conditions and processes we consider desirable have become a reality, and to provide the basis for a discussion on whether the goals we set were, in fact, really desirable ones.

A negligible part of state investment in technology goes to the area of people and computerization. The really large appropriations go to purely technological development for both civil and military purposes. But it is still people who will use this technology and this technology is being developed to serve people.

There are various parts one can play vis-à-vis research; not simply the role of researcher or the user of research findings. One can be an initiator of research, by ensuring that research is started at one's own place of work, one can join a group that works with researchers, or one may be able to conduct some research as part of one's own education/further training.

Let us examine here the need for research and define some of the sub-areas for future research. This discussion is based on an assignment undertaken in 1981–82 on behalf of the National Board of Universities and Colleges (the UHÄ Report on Personnel and Working Life Issues, 1982, p. 19). It is an overview of areas where research has been carried out, research is in progress and where research is needed.

Although some research has been carried out on the following questions, research and development work in the university, for example, has so far largely concentrated on the technical development of computer-based information systems. Research aimed at systems technology design which integrates economic, sociological, psychosocial and medical/ergonomic knowledge on the opportunities and risks that this technology involves is a direct consequence of research conducted in the areas outlined below. Therefore, research that applies the principles of behavioural science and organizational theory, and where the applications of computer technology, both hardware and software, emerge from an interdisciplinary dialogue, should be developed.

Knowledge and research on issues related to computerization and working life should help prevent any undesirable trends from the viewpoint of the individual and the community, should apply the positive effects of computer technology and its uses, and should provide the basic data needed to place limits on the use of technology in society.

1 *Research that analyzes the interplay between computerization and working conditions.* Research in this area aims to identify current lines of development and to produce knowledge that will be used to structure future professional roles. Examples of important sub-issues are the consequences of computer technology and the opportunities it offers for:

- the design and content of jobs;
- the opportunities for personal development in work;
- management;
- social relationships;
- influence on one's own work situation;
- working hours;
- the physical and ergonomic design of the workplace.

Research in this area can provide knowledge about both the short- and long-term effects of technology on working life. Particular consideration should be given to processes and conditions related to the concepts of motivation, cognition and emotion.

2 *Research that analyzes the interplay between computer technology, working life, and the conditions of life outside work* plays an important part in giving us a total picture of the effects of technology on life conditions in general, and this knowledge should make an important contribution towards channelling future developments in a positive direction. It is, therefore,

important to make a detailed study of the impact of computer technology on family life, housing, cultural and social involvement and people's health and well-being. The behavioural sciences (psychology, sociology, medicine) are all needed both in the investigation and in the more normative development of knowledge.

3 *Research on the use of computer technology for the control and democratic development of working life and the community* has become an increasingly strategic area in relation to co-determination. This research covers:

- the analysis of decision-making processes and influence in systems development;
- the analysis and design of different forms of computerized control models for the production of goods and services;
- the analysis and design of information systems to help various interested parties, including trade unions, to assess the potential of a company or a government agency;
- the development of models that make it possible to estimate the total costs of computerization, models that consider both the subjective costs/investments for the individual and the costs/investment for society, government and the local authority.

4 *Socio-structural analyses of computer-based information systems,* including theoretical analyses at the national level, primarily concentrating on the following relationships:

computer technology ⟷ society's standards and system of values ⟷ the labour market and companies

Here, the analysis may include issues related to:

- computer technology and employment levels;
- computer technology and regional development;
- computer technology and the structure of occupations/qualifications.

5 *Educational research aimed at analyzing the role of training*—both company and state-financed—*in the process of computerization:*

- analyses of the function, content, and organization of training/further training in relation to computerization;
- analyses of the learning processes and the development of knowledge;

- an analysis of the long-term consequences of changes in the field of education and learning;
- analyses of the computer as an instrument in education.

2. Research on computer technology, working life, general life conditions

We shall develop the area of point 2 above, since it is related to one of the sub-areas in the RAM:2 Project. There, work was based on my memorandum to the Swedish Work Environment Fund's programme and drafting committee for psychological and social problems in working life (Bradley, 1982).

Computerized information ⟷ Psychosocial factors in the work environment ⟷ Living conditions outside work systems

The above diagram shows the interplay of these three factors. The broad problem area is related to the theoretical model of the relationship of technology, work and living conditions in general developed in the RAM Project and reproduced in Chapter 3. This area was studied in our project (see Chapters 4 and 8), but it is a very broad area, containing different sets of problems which can be illustrated in different projects and by using different approaches, methods, theories, etc.

Research into this interplay is important if we wish to get a total picture of the impact of computer technology on living conditions. It is important to shed light upon issues such as: what aspects of the psychosocial work environment should be given particular consideration when introducing microcomputers? What applications of microprocessor-based equipment are desirable? What areas in working life and in private life should not be computerized (for ethical reasons or considerations of quality, for example)? (The rest of the analyses apply both to the background of computerization linked to the functions in a job, and to computerization linked to functions in the home.) This research should concentrate on two areas.

1 *The interplay of computer technology, working life, and health*

- Are there particular physical and psychological symptoms that can be related to the use of new technology?
- How can data processing technology be structured to reduce strain and stress at work (such as the effects of the rate at which work is done, the perception of time and its relativity)?
- How can new technology be designed to help bring about

desirable conditions seen in terms of our evaluation patterns (e.g., health, well-being, a positive attitude to life), both at work and outside working hours?

- What long-term effects on health can be identified?
- Does the use of 'intelligent' computerized equipment in offices and robots in industry affect our self-esteem?

2 *The interplay of computer technology, working life, and leisure time/private life/family life*

A. Any changes in the structure of our values (such as the benefits we get from our work and our leisure time) as we apply them to work, leisure time, and family life must be analyzed.

- Are there any changes in the way that different aspects of the individual's life situation contribute to his sense of self-esteem?
- Do differences in value structures develop in groups who are dependent in different ways on the new technology in their work and private lives? What is the basic pattern of values in emerging professional groups?
- Does the spread of microcomputers give rise to processes that alienate or involve people both at work and in private life?
- What other fundamental relationships must be considered? These must be analyzed in a life-cycle perspective.

B. Any changes in social involvement (political and/or trade union activities).

- What changes take place in social involvement when the part played by gainful employment in people's lives is reduced? (Does trade union involvement diminish, and is it replaced by an interest in local government politics?)
- Is there any change in the content of the problem areas that are currently dominant in trade union work and in life in general?
- Do differences develop in social involvement between groups with differences in technical knowledge and different experiences of technology, both in their private and professional lives?
- How can computer technology be used to bring about a deeper and more extensive political and/or trade union involvement?
- Will new forms of collective and individual action emerge in a future computer society?

C. Any changes in the choice of work and leisure activities must be analyzed.

- What changes take place in the objective content and psychological benefits of people's leisure time?
- Are there any changes in the balance between different activities outside working hours?
- What changes are there in the pattern of contact, both qualitative and quantitive, in families and circles of friends when communication technology changes both at work and in private life (e.g., an increase in 'remote' contact at work)?
- What changes are there in interest in education set against entertainment and leisure activities, for example?
- Are there any profound changes in the function of the home for the individual and for society? Will there be any drastic change in the balance between the different functions of the family?
- How can technology be harnessed to enrich and deepen the psychological rewards of 'work-free' time?

3. Views on research strategies

Research into technology should be carried out at each of the different levels of analysis (national, corporate, and individual) and encompass different theories and approaches. However, research must place proper weight upon the interplay of these levels. Factors at the national and corporate level are reflected in the awareness, the emotional life, and the thoughts of the individual. These in their turn affect corporate structures and the structure of society. Knowledge of the effects of this kind of interplay will enable us to make deeper and more extensive analyses and develop new theories. It will also open up different ways of influencing developments and shaping a computer society that may be termed humane in the real sense of the word. Research at universities on the computer society should be extended to encompass a broad platform that makes it easier to conduct interdisciplinary research. At the same time, the benefits of single-discipline research could also be applied. When researching the power and the impact of computer technology, one should bear in mind the complexity of the underlying causes, a complexity that involves a number of intricate questions of scientific theory. The research should also provide basic knowledge about people and the organization, focusing on both the inherent opportunities and the risks. Research may focus on issues that are relevant to the development of our society. But research that can help us solve our present problems is also important.

Research into computerization and working life should be conducted on the basis of a broad perspective of equality with regard to sex, age, nationality and disability. It is important to point out the lack of knowledge and the need for research of various groups representing different interests in the community and in the world of research, and for national organizations for financing research, education information, etc. Some of this work is listed in Appendix 4.

4. Some personal reflections on the problems of research

Our society and our working lives are characterized by an increasing degree of changeability. Flexibility, changeability and rationalization are prestigious terms in working life. How much change are we able to assimilate and how well should we prepare ourselves for what we consider to be inevitable changes? For example, what can upbringing and schooling offer us? Our children will need quite different information and skills than their parents did, and they are facing other social patterns than we did. This means that the generation gap will widen and cause us to feel even more uncertain about the future.

The computerization of functions related to our private lives is just around the corner—and here there is considerable potential for manufacturers in the computer industry. New technology is grafted onto old equipment, while new and more advanced machines and instruments are being developed. Can this imminent change be compared to a cultural revolution? Will we expand the use of 'home computers', and will they make housework even easier, or will microelectronics invade our everyday lives to a point where a 'leap' in development will occur, changing beyond recognition the function of the home for the individual and for society? What changes do we want to see and how contradictory are the goals we shall attempt to attain?

Different groups, divided by sex, age, educational background (poorly educated/highly qualified) face different circumstances, and have different reactions and attitudes. For example, in the case of sex, there is so-called extrinsic and intrinsic equality. The extrinsic aspects of equality are intimately linked to the allocation issue (see below). A crucial objective for extrinsic equality is to attain a fairly even distribution of resources among the different activities, not least in powerful and influential positions. Here, it is important that we pay particular attention to employment, the structures of professions and the divisions of labour, both vertical and lateral.

What quality of life do we want? What factors of our life do we value?

Photo: Göran Assbring/MIRA

Intrinsic equality refers to men and women as a whole and as carriers of separate cultures. The aim here is that the experience, values, lifestyles, attitudes, etc. that are the traditional province of women be given parity with those of men in the different aspects of the life of our society. In this context, it is important to consider the psychosocial aspects of the work environment, the balance between working life and private life, between family life and social life, etc. Will computer technology help the traditional male culture to assimilate a traditional female culture? Or will we see these forces in equilibrium? Technical change and, naturally, computers as well, point up the rational components in human thinking and human actions. We must acknowledge the importance of having both the emotional and rational parts or sides of ourselves expressed in working life and in the community.

A negative development is possible: that efficiency and rationalization will become altogether too dominant in work and in our private lives. But a positive development is also possible: that the computer will be used to give us more time for the functions and tasks that are related to our emotions, such as associating with each other, serving other people, listening, and becoming involved. Computers may also help us to eliminate alienation in many jobs, jobs that were created in the current stage of the automation process, and replace them with meaningful jobs and work tasks that will encourage self-development (irrespective of whether or not these tasks come under the heading of 'work'). At the cognitive level, there is a risk that people will become more 'ignorant' and lose their knowledge, but there is also a chance that computers may be used to stimulate both creativity and imagination.

The outcome depends on our work and our involvement at all levels—at national, corporate and individual levels and involves both questions of allocation and guidelines for action that aim to bring about desirable conditions of life.

Appendix 1. The Main Content of the Indices

The main content of the indices is described here in the form of composite questions related to the questions in the indices. The questions are set out in full in *Work Environment and Display Terminals* (Bradley, 1981, Appendixes 2 and 5).

The S-indices: Psychosocial Work Environment Factors

Interesting work—meaningfulness (SIME)
Do you enjoy your work? Are your work tasks interesting and enjoyable? Do you feel that your work is pointless? Do you feel as though you are simply 'putting in hours at work'?
Interesting work—human resources (SIMR)
Is your work not complex enough? Do you want your job to include a large number of work operations? Does it allow you to use your own knowledge and skills?
Job autonomy (SSJA)
Do you have the opportunity to discuss the way your work is structured and performed? Is there enough self-determination in your work? Is your work too strictly monitored? Are you able to influence your work conditions?
Opportunities for contact with co-workers (SKONT)
Are you too confined to your workplace or too isolated? Do you have sufficient opportunities for contact with your fellow-workers and with other departments in the company?
Cooperation with co-workers (SSAM)
Do you enjoy good cooperation with your co-workers? Do you get on well with your closest co-workers? Do you work well together?

Wage conditions (SLÖN)
Are you underpaid compared to others outside the company or in consideration of the work you perform? Do you feel that wage differences are too great in your company?

Promotion—positive (SBEFP)
Will promotion give you a more interesting job? Is this a positive stimulus for the individual? Will promotion give you more influence over your work?

Promotion—negative (SBEFN)
Are there rumours of intrigue when people are promoted? Does promotion create competitive relationships?

Promotion prospects (SBEFM)
Does the company place sufficient value on long experience/professional skills? Do you have good prospects of advancing to more qualified work tasks? Is there scope in your present position for personal development?

Information (SINFO)
Are you given adequate information on events in the company, changes at the workplace, e.g., rationalization? Are you given adequate information on the conditions that affect your own job, and information on events in other departments?

Intermediate position—the buffer role (SBUFF)
Do you feel that other groups, e.g., the general public/customers, make far too high demands of you? Do other departments or personnel groups within the company make far too high demands?

Work pace/work load (SARBT)
Is the work pace usually too high? Is your work hectic and stressed? Is your department undermanned? Does the demand for efficiency dominate? Does the job generate pscychological strain?

Irregular work pace/work load (SOJA)
Is the pace of work too uneven, too dependent upon other people? Are you bothered by too high work peaks at certain times of the year?

The status of the work (SAANS)
Does your job have high status in the company and among your friends?

Attentiveness/responsibility (SUPPA)
Does the work require excessive attentiveness and precision, and does it involve too much responsibility?

Efficiency (SEFF)
Can you be as efficient as you yourself would wish in consideration of the people in your own department and in other departments that you depend upon, and in consideration of the essential equipment you have?

The F-indices (Physio-Ergonomic Work Environment Factors)

Work premises (FA)
Are you satisfied with your work premises? Are the premises well-suited to the work performed there? Are you troubled by the lighting conditions or by noise at your workplace?
Temperature/air (FT)
Are you satisfied with the temperature and ventilation of your workplace?
Ergonomic factors (FE)
Are the furniture, the aids, the work tools and other equipment you use suitable for the work and do they make the work easier to perform? Is your work position comfortable? Are lighting conditions suitable for the work you perform?
Odours (FL)
Are you troubled by smells or by your co-workers' smoking?

The DEM-indices: Democracy at Work

Work democracy (DEMA)
How much influence do you have on the planning of your own work, regulating your work tempo, your choice of work method? Do you have enough influence on your own work and working conditions?
Company democracy (DEMF)
How much influence do you have on the planning and implementation of rationalization, the choice of your immediate co-workers (not employment), the employment of personnel at the company, questions that are of general interest to the company?
Company democracy—preferences (DEMFÖ)
Do you consider that there should be a change in your influence in terms of the planning and implementation of rationalization, appointments to positions and matters that are of general interest to the company?
Work/leisure time democracy (DEMAL)
How much influence do you have on the scheduling of breaks, holidays and working hours (within the framework of 'flexitime')?
Work/leisure time democracy—preferences (DEMALÖ)
Do you consider that there should be a change in the amount of influence you have on the scheduling of holidays, working hours and breaks?

The T-indices: Effects of the Terminal System

Interesting work/job autonomy—T(TISJA)
Has the introduction of the terminal system led to any changes (increase or decrease) in the following: varied work assignments, the amount of stimulus and tension, the amount of interesting job content, the number of different operations that the work consists of, opportunities to apply your own ideas, knowledge and skills?
Errors in work—T(TFEL)
Has the introduction of the terminal system led to changes (increase or decrease) in the following: the risk of making mistakes, anxiety about making mistakes and the consequences of errors, the opportunities to correct errors?
Opportunities for contact with co-workers/cooperation—T(TKONTSAM)
Has the terminal system had any effect on the unity/cooperation of yourself and your immediate co-workers, and among employees throughout the department? Has there been any effect on the opportunities for contact with co-workers who have similar types of work, with personnel in your department, with your immediate co-workers, or with other departments within the company? Has the opportunity for contact with your immediate superiors and your ability to get on with them been affected by the introduction of the terminal system?
Intermediate position—the buffer role—T/D (TDBUFF)
Has the introduction of the terminal or data processing system brought about changes (increase or decrease) in the following: demands from other departments or personnel groups within the company, simultaneous demands from both superiors and subordinates, demands from other associated groups?
Influence—work democracy—T(TINFLAD)
Has the introduction of the terminal system led to any change (increase of decrease) in influence over the following: your own work pace, planning your work, choice of work methods, scheduling breaks?
Influence—company democracy—T(TINFLED)
Has the introduction of the terminal system had any affect on or influence over the following: appointing people to positions, choice of your immediate co-workers, planning and implementing rationalization measures?
Irregular work pace—T(TOJA)
Has the introduction of the terminal system led to a decrease or increase in the unevenness of the work pace and the degree to which your work pace is dictated by others?
Attentiveness/responsibility—T(TUPPA)
Has the introduction of the terminal system had an effect upon your

responsibility at work and the requirement for your attentiveness and precision?

Wages—T/D (TDLÖN)

Has the introduction of the terminal or data processing system otherwise affected the salary levels at the company, your own pay and your own prospects of pay increases?

Working hours—T/D (TARBTID)

Has computerization meant that you have working hours that suit your home and family life and leisure activities? Do you now have more secure working hours and less overtime work?

Interest in training—requirements—T(TUTIK)

Has the introduction of the terminal system had any effect on your own interest in training or upon the company's demands for training/further training?

Ergonomic factors—T(TERGO)

Has the introduction of the terminal system led to improvements or deterioration in the following: furniture, lighting, work positions, the modification of the work premises to the work performed in them and the mechanical equipment?

Temperature/air—T(TTEMPL)

Has the introduction of the terminal system led to changes in terms of temperature and ventilation?

Health hazards—T(THÄLSOR)

Have conditions related to the introduction of terminals reduced the risk of mental and physical complaints?

The D-indices: Other Effects of the Computer System

Interesting work/job autonomy—D(DISJA)

Have there been any changes (increase or decrease) in the following as a result of the other effects of the computer system: variety at work, the degree of stimulus and tension and the amount of interesting job content, the opportunity to use ideas, knowledge and skills, autonomy/freedom of action in the work, the status of the job, the number of different operations included in the work?

Errors in the work—D(DFEL)

Has the introduction of the computer system otherwise led to change (increase or decrease) in the following: risks of making mistakes, anxiety about making mistakes, opportunity to correct errors?

Contact opportunities—D(DKONT)

Has the computer system otherwise affected your opportunities for contact with your immediate co-workers, with your immediate superior, with co-workers with similar work tasks/position, with the other personnel in the same department?

Cooperation—D(DSAM)
Has the introduction of the computer system otherwise affected the cooperation and sense of affinity between yourself and your immediate co-workers, with your immediate superiors, among employees in the department, among co-workers with similar work assignments/ positions?

Influence—work democracy—D(DINFLAD)
Has the introduction of the computer system otherwise led to any change (increase or decrease) in influence over the following: your own work pace, planning your work, choice of work methods?

Influence—company democracy—D(DINFLFD)
Has the introduction of the computer system otherwise had any effect on or influence over the following: appointing people to positions, choice of your immediate co-workers, planning and implementing rationalization measures, influence in matters of general interest to the company?

Work pace/work loads—D(DATAB)
Has the introduction of the computer system led to a general decrease or increase in the dependence on other people of the work pace, unevenness in work pace, stress of work, the work pace's dependence on machines, equipment, etc?

Interest in education—requirements—D(DUTIK)
Has the introduction of the computer system had any general effect on your own interest in training or upon the company's requirements for training/further training?

Ergonomic—D(ERGO)
Has the introduction of the computer system otherwise led to improvements or deterioration in the following: lighting conditions, furniture, the work position, the modification of the work premises to suit the work that is performed there?

Temperature/air—D(DTEMPL)
Has the introduction of the computer system led to changes of temperature and ventilation?

Health risks—D(DHÄLSOR)
Have conditions related to the introduction of terminals reduced the risk of mental and physical complaints?

The HA-indices: Health Problems at Work

Nervous complaints—A(HAN)
Have you been troubled by depression, nervousness (worry, anxiety), restlessness, irritation, fatigue, irregular heartbeat or periods in which you have found it difficult to summon the energy to achieve anything at work?

Eye complaints—A(HÄO)
Have you ever been troubled by fatigue, burning sensations or other eye complaints at work?
General pain—A(HAV)
Have you been troubled by pain or fatigue in your shoulders and neck, had hand or finger joint problems, swollen feet and/or legs or suffered from back complaints at work?
Stomach complaints—A(HAM)
Have you been troubled by stomach/intestinal complaints, nausea or diarrohoea at work?
Headaches, etc.—A(HAH)
Have you been troubled by migraine or headaches at work?

The HF-indices: Health Problems during Leisure Time

Nervous complaints—F(HFN)
Have you been troubled by depression, nervousness (worry, anxiety), restlessness, irritation, fatigue, irregular heartbeat or periods in which you have found it difficult to summon the energy to achieve anything in your leisure time?
Eye complaints—A(HFÖ)
Have you been troubled by fatigue, burning sensations or other eye complaints in your leisure time?
General pain——A(HFV)
Have you been troubled by pain or fatigue in your shoulders and neck, had hand or finger joint problems, swollen feet and/or legs or suffered from back complaints in your leisure time?
Stomach complaints—A(HFM)
Have you been troubled by stomach/intestinal complaints, nausea or diarrhoea in your leisure time?
Headaches etc.—A(HEH)
Have you been troubled by migraine or headaches at work?
Sleeping problems—F(HFS)
Have you had difficulties in sleeping or used sleeping tablets or tranquilizers during the last twelve months?

The AF-indices: The Effects of Work on Leisure Time/Family Life

The effects of work on leisure time/family life (AF)
Do you feel that your current position makes too great inroads into your home and family life or your leisure time activities? Do you do too much overtime work? Is it difficult to plan your leisure time because of an unpredictable work schedule?

The DF-indices: The Effects of Computerization on Opportunities for Leisure Activities

Leisure activities—D(DFSYM)
Do you consider that computerization at the company directly or indirectly had an effect on your opportunities to go to the cinema, the theatre, dance restaurants, your opportunities to read, take part in open-air pursuits, take exercise, take part in sports or go shopping or window-shopping?
Contact—D(FKM)
Do you consider that computerization at the company has directly or indirectly reduced your opportunities for a home and family life, association with friends or your chances of studying or attending courses of further education?

The V-indices: The Weight Attached to Specific Work Environment Factors and the General Attitude to Work

Weighting—external (VY)
What factors are important in helping you enjoy your job: good working hours, good physical work environment, a job with a calm and even pace of work, a job that does not impinge on family life/leisure time?
Weighting—internal (VI)
What factors are important for your enjoyment of your job: having varied work, having a job that gives you the opportunity for personal development, having influence on decisions that apply to your own work and your own working condition, having contact with people at work, having an interesting job?
Weighting—development (VU)
What factors are important for your enjoyment of your job? Having good promotion prospects, good opportunities for training/further training, having a job with scope for personal development?
Interest in promotion (VB)
Are you interested in applying for promotion, do you feel the need to advance, are you interested in further education to prepare yourself for other work tasks within your department?
(Low values of these V indices mean that you consider the work environment aspect to be important.)

Appendix 2.
Reports and books from the RAM Project: computer technology and working life, studies of the effects of rationalization (in particular computerization) on the work environment of salaried employees

A detailed description of the findings of the Post Giro Centre is given in RAM reports nos. 1–6 (ISSN 0349–2915).
The preliminary findings from the Post Giro project are reproduced in Bradley, G. (1977) *Computer Technology, Working Life and Communication* (Stockholm, Swedish Council for Future-Oriented Research, Liber Förlag) and Bradley, G. (1979) *Occupational Roles and the Quality of Life* (Stockholm: Wahlström and Wikstrand, ISBN 91–46–13409–3). The results from the Trygg-Hansa survey are published in Bradley, G. (1981) *The Work Environment and Display Terminals* (Stockholm, Swedish Work Environment Fund/Liber/Allmänna Förlaget, ISBN 91–38–06353–0) and in RAM reports nos. 8–10 (ISSN 0349–2915).

Documentation from the RAM Project

The research reports are published in limited editions. In order that the results of the RAM Project should reach a wider readership, the material has been published in a series of reports that have been distributed to large libraries in Sweden and the other Nordic countries, etc. Some of these reports are still available in limited numbers and may be ordered from The Swedish Work Environment Fund. Essays and papers from the project, and articles published in Swedish and foreign journals are not included here.

Reports from the RAM Project (ISSN 0349–2915; in Swedish unless otherwise stated)

RAM report no. 1
Bradley, G. (1976) *Work Environment and Computerization* (University of Stockholm). Qualitative analyses of the relationship between administrative rationalization with EDP and the psychosocial aspects of the work environment.

RAM report no. 2
Bradley, G. and Nilsson, I. (1976) *Work Environment and Computerization* (University of Stockholm). Quantitative analyses of the relationship between administrative rationalization with EDP and the psychosocial aspects of the work environment.

RAM report no. 3
Bradley, G. and Nilsson, I. (1976) Appendix in tabular form for RAM report no. 2, *Work Environment and Computerization.*

RAM report no. 4
Bradley, G. and Nilsson, I. (1977) *Work Environment and Computerization* (University of Stockholm). A summary of reports 1 and 2.

RAM report no. 5
Bradley, G. (1979) *Computer Technology and Working Life* (University of Stockholm). Memorandum for the Nordiska Akademikerrådet conference, 1979, 'Surveying the Work Environment of the Long-Range Educated'.

RAM report no. 6
Bradley, G. (1979) *The Interplay between Computer Technology — Work Environment — Physical and Psychological Symptoms* (University of Stockholm).

RAM report no. 7
Bradley, G. (1979) *A Collection of Articles* (University of Stockholm). A reprint of articles published between 1974 and 1978 dealing with the psychosocial work environment.

RAM report no. 8
Bradley, G. (1980) *Computers and Work Environment from a Psychosocial Point of View* (University of Stockholm). Report to the TCO Conference, 1980, on 'Computer Development: Control or Be Controlled'.

RAM report no. 9 (available in English)
Bradley, G. (1980) *Computerization and Working Life from a Psychosocial Point of View* (University of Stockholm).

RAM report no. 10
Bjerin, A. and Martens, P. (1981) *Effects of the Work Environment on Leisure, Family and Health* (University of Stockholm).

RAM report no. 11 (available in English)
Bradley, G. (1983) *Computers and Their Psychosocial Impact, Especially with Regard to Women.* A report presented at the International Women's Seminar in Saltzburg, 1982.

RAM report no. 12
Martens, P. (1984) *The Relationship between Working Life and Family Life.*

Books from the RAM Project (in Swedish unless otherwise stated)

Bradley, G. (1977) *Computer Technology, Working Life and Communication* (Stockholm: Liber Förlag. Forskningsrådsnämnden) Swedish Committee for Future-oriented Research.

Bradley, G. (1979) *Occupational Roles and the Quality of Life* (Stockholm: Wahlström and Widstrand).

Bradley, G. (1981) *The Work Environment and Display Terminals: Analyses of Terminal Systems — Psychosocial Aspects of the Work Environment* (Stockholm, Swedish Work Environment Fund/Liber/ Allmänna Förlaget).

Bradley, G., Bergström, C. and Sundberg, L. (1984) *The Working Roles of Secretaries and Word Processing Systems* (Stockholm: Swedish Work Environment Fund). Changes in the office environment.

Bradley, G., Bergström, C. and Lindeberg, S. (1988) *The Role of Engineers and the Future Computer Technology* (Stockholm: Carlsson).

Appendix 3. A sample checklist for evaluating the working environment of computer systems, both planned and in use

This checklist may be used in the evaluation of a projected computer system or a system which is already online, to give a rapid assessment of which aspects of the work and working conditions are to be improved and developed. The key words in the left-hand column may be read first. The right-hand column contains examples of the areas to be selected for analysis and action.

Aspect of work and the working environment	Examples:
Work content	The value of the work to the company. The value of the work to society. The value/significance of the work to the individual. The number of work operations in the job. Opportunities to apply one's own knowledge and ideas (professional skills). The opportunity to 'follow through on assignments'. Monotony, repetitive work operations. The opportunity to relate your work to the

	work of the organization as a whole. Supervision/monitoring of the job. Being confined by certain work routines. Being confined to a work station. Variety in the job. The degree of complexity in the work. The qualifications required for the job. The degree of independence.
Job autonomy	Influence on the way the job is performed (work method). Influence on the work design. Influence on the work planning. Influence on the pace of work. Influence on working hours. Influence on rest and work break scheduling. Influence on the choice of co-workers. Influence on the employment of personnel. Influence on the planning and implementation of rationalization measures.
Security/safety	Security of employment. Security and development opportunities in your present position. Transfers. Reductions in personnel. The opportunity to take another position (at the same level). Recruitment.
Contact/cooperation with co-workers	Contact/cooperation with other people at work immediate superiors immediate subordinates closest co-workers other personnel in the department other departments the general public/customers. Communication between people and computers compared with communication between people. Confinement to the workplace. Isolation (physical and psychological). Too much/too little contact with others. Knowledge of other people's work. Group work/individual work.

Work load	The pace of work (manning levels).
	An irregular work pace.
	The extent to which machines/equipment dictate your pace of work.
	The extent to which other people dictate your pace of work (e.g., co-workers, other departments).
	The opportunity to take work breaks.
	The need for attentiveness in the work.
	The need for precision.
	Replaceability
	  can you be replaced at work (in case of sickness, etc.)
	  the difficulty of finding a replacement.
	Perception of personal efficiency at work.

Responsibility	Different aspects of responsibility (compare with job autonomy).
Working hours	The scope of working hours (full/part-time). The scheduling of working hours (flexible working hours, shift work). Overtime work, unscheduled working hours.
The effects on leisure time/family life	Home and family life. Further training/courses. Trade union/political activities. Hobbies. Taking exercise. Watching TV, reading. Spending time with friends.
Promotion/ development	The opportunity to develop in the job. Prospects of traditional professional advancement. Prospects of getting more qualified work at the same level. The possibility of 'declining' promotion — psychologically.

Pay and income	The design of the pay system.
	Pay according to the work performed
	  other groups in the company
	  others with similar work in the company.

Relationship to the	Distance (psychological and physical).

company (in general)	Commitment. Conflicting demands superior/subordinate other departments the general public/customers private life Information (upward, downward, lateral) quantitative qualitative
Status	Status vis-à-vis other departments. Prestige, status vis-à-vis fellow-workers.
Educational aspects	Changes in qualification requirements (higher, lower). In-house training. Training for new work tasks. Training in the EDP system. Interest in acquiring more knowledge.
Physio-ergonomic conditions, health aspects	Are the work premises suited to the nature of the work? Work position. Furniture Work tools/equipment (mechanical equipment). Lighting. Sound and noise. Temperature. Smoking. Risks of physical and psychological complaints.
Work organization	The operational goals. Formal structure. Informal structure. The design of the rewards system. The management system. Merging departments (lateral changes). Vertical changes (transferring responsibility, the managerial function). The representation of the sexes (positive action in recruitment and appointments). The age profile.
Technical/mechanical aspects	Information on processing stoppages. Manual errors worry about making mistakes

the possibility of correcting errors
the consequences of errors.

Response times.

Accessibility.

Utilization of technical capacity.

Problems of confidentiality.

Appendix 4. Changes in working life – Research priorities

Listed below are a number of published and unpublished reports by Gunilla Bradley identifying research needs and priorities related to changes in working life:

- Project outlines in the field of *Sex Roles and Working Life* Stockholm, The Swedish Council for Personnel Administration (1970).
- Chapters on proposed research areas in the book *The Working Environment and Salaried Employees* (Bradley, Börjeson and Lundgren, 1974).
- A project programme for The Work Environment Fund (1973) took up four broad research areas linked to computerization:

  Employment and computerization
  Occupational structures and computerization
  Economics and computerization
  The working environment and computerization.

  Studies of the Effects of Rationalization on the Working Environment and Working Conditions for Some Groups of Salaried Employees, a project programme for The Work Environment Fund (1974).
- Chapter 5 of *Computer Technology, Working Life and Communication*, the Delegation for Long-Term Motivated Research, The Co-ordinating Board of the Swedish Research Councils (1977).
- *Computer Technology and Working Life*, a project programme for The Work Environment Fund (1979).
- *Computer Technology and Working Life* a report to the Council of Academies 1979 conference (RAM report no. 5).
- A memorandum to The Work Environment Fund's working group for planning research in the psychosocial field (1982).
- A memorandum to the committee in the National Board of Universities and Colleges on urgent research in the field of computer technology.

Appendix 5. Labour Relations in Sweden

Sweden is regarded as having a highly developed labour market. It has attracted attention because of its strong business sector, its high level of trade union membership, peaceful cooperation between employers and trade unions, and closely regulated labour relations.

Some basic facts

Sweden has 8.4 million inhabitants of whom no fewer than 4.4 million—more than 52% of the population—belong to the labour force.

Just over a century ago, Sweden was still an agrarian country. At about that time, there began a process of industrialization which has continued through our own times. In the past 20–30 years, however, the largest growth in jobs has been in the public sector. Private enterprise accounts for 63% of employment in Sweden today, the public sector 37%; comparable figures for 1965 were 79% and 21% respectively.

Sweden is usually regarded as the promised land of special-interest organizations. A very large proportion of eligible employees belong to trade unions—about 85%—and to numerous organizations in other fields. Private ownership of the business sector is nevertheless larger than many people realize—also around 85%.

During the 20th century, the average real industrial wage has roughly quintupled, while at the same time the number of working hours per year has fallen by more than 40%, from 3,100 to 1,800. The following table shows annual labour costs for an industrial worker since 1900, converted into 1985 prices:

1900	SEK[1]	26,000
1920	SEK	31,000
1940	SEK	42,000
1960	SEK	72,000
1986	SEK	140,000

[1] SEK 1 (Swedish krona) = US$0.17 or £STG 0.10

Of the SEK 157,100 paid in 1986, about SEK 110,400 consisted of wages (SEK 39,400 of this was withheld by the employer for income tax payments) and SEK 46,700 comprised social welfare contributions paid by the employer to the government on behalf of each worker. The corresponding figures for white-collar workers in the business sector were a total labour cost of SEK 219,300: gross salary of SEK 150,500 (of which SEK 61,455 was withheld for income taxes) and social welfare contributions of SEK 68,800.

As indicated above, employers pay social welfare contributions to the public sector, in addition to meeting payrolls. As of 1987 these fees comprised 43.2% of annual wages in the case of blue-collar workers (1973: 18.1%) and 46.3% of the salaries of white-collar workers. The largest of these items are the fees which finance the national supplementary pension system (10.2%), the health insurance system (9.3%), the basic pension (9.45%) and—in the case of white-collar workers—private pension systems (8.25%).

Historical background

The first union in Sweden was formed in 1846 when the Typographical Association was established in Stockholm, but the concept of organized labour in the modern sense emerged only after the industrial revolution began during the 1870s and 1880s. In the 1880s the first national trade unions came into being, most of them based on the craft principle. In 1898 a number of unions joined forces to form a central organization known as the Swedish Trade Union Confederation (*Landsorganisationen i Sverige, LO*). By this time 27 national labour unions existed, but not all of them joined LO immediately.

A few years later, in 1902, the employers formed the Swedish Employers' Confederation (*Svenska arbetsgivareföreningen, SAF*). In 1906 LO and SAF began to realize they had to recognize each other's rights, and they signed the so-called December Compromise.

In this agreement, SAF formally recognized that the workers were entitled to form organizations and to negotiate. LO recognized the right of the employers to hire and fire workers freely, to direct and allot work and to hire any workers they pleased (i.e. no closed shop clauses were allowed; this is still the case today).

The General Strike of 1909 brought a halt to this early period of cooperation, and for many years conditions on the labour market were unclear, with the employers holding the stronger position. During the 1920s, new labour legislation began to be passed. Of particular importance were the Collective Bargaining Act and the Labour Court Act, both adopted in 1928.

The 1930s witnessed the beginnings of labour-management cooperation on the basis of greater mutual trust. In 1938 SAF and LO concluded the Saltsjöbaden Agreement, a major event at that time (Saltsjöbaden is a seaside resort near Stockholm). This basic agreement regulated the mutual relations between labour and management with regard to negotiations and industrial action.

Meanwhile, in the early 1930s, white-collar workers had begun to organize themselves into trade unions completely independent of LO, in many cases by expanding or adapting professional and social organizations formed long before then.

The Central Organization of Salaried Employees (*Tjänstemännens centralorganisation, TCO*) was formed in 1944 as the result of a merger between two central white-collar confederations, one representing private-sector employees (founded in 1931), the other civil servants (founded in 1941).

The Swedish Confederation of Professional Associations (*Sveriges Akademikers Centralorganisation, SACO*) was formed in 1947. At first the criterion for membership was a university education or the equivalent, but its recruitment principles were later modified somewhat as a result of changes on the labour market. In 1975 SACO merged with the National Federation of Civil Servants (*Statstjänstemännens riksförbund, SR*) to form a joint confederation called SACO/SR.

During the 1940s, a number of basic agreements aside from the Saltsjöbaden pact were signed. They concerned workers' protection, vocational training, works councils (including both employer and union representatives) and time-and-motion studies. During the 1950s, similar agreements were concluded in the white-collar area as the organizations in this sector gained strength.

Because of these agreements and the cooperative "spirit of Saltsjöbaden" between labour and management that accompanied them, the number of strikes on the Swedish labour market remained low for several decades, except for a prolonged conflict in the engineering industry in 1945. The relative lack of industrial action was a strong contributing factor to Sweden's growing prosperity during the 1950s and 1960s. A "Swedish Model" took shape, characterized by centralized collective bargaining, an active government labour market policy and very few labour disputes. Companies such as Alfa-Laval, ASEA, Atlas Copco, Electrolux, Ericsson and many others grew strong.

During the 1970s Sweden, like other countries, encountered major economic setbacks which continued into the 1980s. A wide range of labour legislation was adopted, with the 1977 Act on Employee Participation in Decision-Making (also called the Co-Determination Act or by its Swedish acronym MBL) attracting the most attention.

At the confederation level, there is a gap between the views of employers and unions on quite a number of issues, particularly in regard to the employee (or wage-earner) investment funds and on pay policy. There is, nevertheless, continued close labour-management cooperation in most areas, and since a major industrial conflict in 1980 there has been little industrial action in Swedish industry.

Trade unions

Unlike most countries, Swedish has separate trade unions for blue-collar and white-collar workers. The three confederations mentioned above dominate the picture—one for blue-collar workers and two for the white-collar sector.

By far the largest confederation of trade unions is LO. With its 2.3 million members (1.8 million of these are in work), LO covers about 90% of all blue-collar workers. LO has always cooperated closely with the Social Democratic Party, which ruled Sweden during the period 1932–1976 and returned to power in 1982.

LO today comprises 24 nationwide trade unions with 1,200 branches and roughly 12,000 local branches or sections. The largest national unions are the rapidly growing Swedish Municipal Workers' Union with 640,000 members, the Metal Workers' Union with 460,000, the State Employees' Union with 210,000, the Commercial Employees' Union with 160,000 and the Building Workers' Union with 150,000 members. The LO unions and the confederation itself have a total of about SEK 5.5 billion in their conflict funds.

The rapid growth of the Swedish trade union movement can be seen in the following table:

Membership of LO 1900–1980, in millions

1900	0.04
1920	0.28
1940	0.97
1960	1.49
1980	2.13
1987	2.28

In the white-collar area, the dominant confederation is TCO. With 1.1 million active members, TCO covers about 75% of all white-collar workers, which is a high figure in international terms. The largest of the 20 national unions affiliated with TCO is the Swedish Union of Clerical and Technical Employees in Industry with 280,000 members. The second-largest is the Swedish Union of Municipal Employees with 170,000, followed by the Federation of Civil Servants with 121,000 and the Union of Commercial Salaried Employees with

113,000. The TCO unions have about SEK 3 billion in their conflict funds.

The third major union confederation, as mentioned above, is SACO/SR, which now has 280,000 members. Most of these are employed within the public sector. The largest of the 25 national unions affiliated with SACO/SR are the National Union of Teachers in Sweden with some 45,000 members and the Swedish Association of Graduate Engineers with 45,000.

In addition to LO, TCO and SACO/SR, there are some smaller unaffiliated trade unions. They have little impact on the labour market as a whole.

Because TCO, unlike LO, does not take part in pay talks, three distinctive negotiating bodies have been formed in the white-collar area. Within the private sector, seven unions affiliated with TCO and 14 affiliated with SACO/SR negotiate jointly with employers through the Federation of Salaried Employees in Industry and Services (*Privattjänstemannakartellen, PTK*). PTK was formed in 1973 and negotiates on behalf of about 500,000 union members, but is only empowered to sign recommendation agreements on behalf of the affiliated organizations.

There are corresponding negotiating bodies for the local and national government sectors. TCO's Section for State Employees, TCO-S, and the Federation of Salaried Local Government Employees *(Kommunaltjänstemannakartellen, KTK)* conclude agreements that are binding on member associations. Participation in KTK is also open to non-TCO unions with members in local government service. SACO/SR has corresponding cartels which negotiate for the local and national government sectors.

LO and to some extent TCO are organized on the industrial principle, which means that employees belong to national unions according to the economic sector in which they work, rather than by their profession. But some of the TCO unions as well as the member organizations within SACO/SR are structured according to the craft union principle.

Employers

The dominant employer organization in the private sector is SAF, which has 42,000 member companies employing 1.25 million people, 60% of whom are LO members. Most SAF member companies are small (more than half of them have five employees or fewer). At one time SAF was primarily a negotiating body, but in recent years it has followed the example of LO and TCO and has come to devote equal energy to influencing public opinion. SAF has about SEK 2.8 billion in its insurance fund.

The largest of SAF's 35 member federations are the Swedish Engineering Employers' Association, whose member firms employ 325,000 people, the Commercial Employers' Association with companies employing 200,000, the SAF General Group with 124,000 and the Employers' Federation of the Swedish Construction Industry with 80,000.

Aside from SAF, there are separate employer organizations representing banks and publishers of periodicals. The Cooperative Employers' Association (*Kooperationens Förhandlingsorganisation, KFO*), represents companies which have a total of 100,000 employees. The Employers' Association SFO (*Arbetsgivarföreningen SFO*) represents companies and organizations with 125,000 employees.

In 1966, employees in the public sector were granted a major expansion of their right to negotiate and were empowered to go on strike and carry out related industrial action. The number of people working for the municipal governments and county councils in particular has grown very much in recent years. The National Agency for Government Employers (*Statens Arbetsgivarverk, SAV*) represents the State in its role as employer, negotiating with about 580,000 employees (including not only national civil servants but also municipally employed teachers). The Swedish Association of Local Authorities (*Svenska Kommunförbundet*) negotiates with 530,000 employees of Sweden's 284 municipal governments; the Federation of County Councils (*Landstingsförbundet*) negotiates with some 410,000 employees of the country's 23 county councils.

Collective bargaining

Over the years Sweden has avoided pursuing a State incomes policy. The employer organizations and the trade union movement are regarded as strong enough to reach their own agreements on pay and other working conditions. These conditions have normally been regulated by nationwide collective bargaining agreements. But due to government taxation policies, the inflation rate and the shrinking space for real pay increases, there has been a growing degree of consultation with the Cabinet on various pay and income issues.

The most important function of collective bargaining contracts is to preserve labour peace. If no such contract is in force, industrial action—strikes and lockouts—is permitted. Collective bargaining contracts also have a norm-creating function, i.e. a company and its employees are not allowed to sign agreements stipulating conditions which are poorer than those fixed by the applicable collective contract.

Labour legislation

It was characteristic of the 1970s in Sweden that labour market issues that were previously regulated by collective bargaining agreements increasingly came to be regulated by legislation. Once fundamental social rights had essentially been achieved, politicians became increasingly interested in labour market issues. The employee side believed that the time was ripe to raise the standard of the labour market regulation provided by laws.

In 1971 the LO Congress, which meets every five years, adopted a wide-ranging programme calling for greater employee participation in decision-making and other changes in labour law. TCO likewise adopted a programme which called for greater co-determination. As a result of these demands, a large number of new labour laws were passed between 1973 and 1977. Some of the more important ones are described below.

First, however, a few words about the Labour Court (*Arbetsdomstolen*), which was founded in 1928 and comprises the final—and in most cases the only—forum for settling legal disputes on labour issues. Its members include representatives from the State, the employers and the trade unions. In major cases, seven members sit on the Court—three representing the State and two representing labour and management respectively. Since 1974 the Court has heard 200–300 cases annually, after previously hearing substantially fewer cases for many years.

In 1974 the important *Security of Employment Act* went into effect and significantly increased the influence of employees and trade unions. Dismissal of an employee must be based on so-called objective grounds. Such grounds do not exist if it can reasonably be demanded that the employee be transferred to a different position in the same company. Older employees are entitled to longer periods of notice than younger ones: those under age 25 get one month and those 45 or over get six months. Former employees enjoy priority when a company is again hiring.

After criticism from employers and others, a revised Security of Employment Act was adopted in 1982. It re-introduced the option to hire people on a trial basis for up to six months, among other things in order to encourage the creation of new jobs.

The 1974 *Act Concerning the Status of Shop Stewards* entitles trade union representatives to perform their union duties on paid working time. Local branches decide which people are to be regarded as union officials, and the local branches enjoy priority when it comes to deciding what time these shop stewards may devote to union activities at the workplace.

A 1975 law increased opportunities for employees to be granted *leaves of absence from work to pursue studies* for long or short periods.

The *Working Environment Act* of 1978 replaced an earlier workers' protection law and increased the rights of trade unions to help improve the working environment. The concept of the working environment now encompasses work systems, working hours, and adaptation of work to human factors, both physical and psychosocial. Safety stewards (delegates) were granted the right to halt dangerous work under certain circumstances. The SAF-LO-PTK working environment agreement of 1976 regulates the tasks and responsibilities of the safety committees required at larger workplaces to supervise their occupational health and safety programs.

The 1980 *Act on Equality between Men and Women* forbids an employer from discriminating against an employee on account of sex. Men and women should enjoy equal opportunities for employment, training, promotion and on-the-job development. They should receive equal pay for work of equal value.

The *Vacation Act* of 1978 increased the statutory minimum paid vacation from four to five weeks.

Employee participation

One of the most complex labour market issues in Sweden during the 1970s was employee participation in decision-making at the workplace. A long series of changes and innovations took place. Three different levels of participation have generally been mentioned: shop floor participation, company participation and financial participation.

Shop floor participation refers to the ultimate purpose of co-determination reforms, i.e. to enhance the individual's influence over his or her own working situation. Working systems are one area in which companies and government agencies have shown great interest.

Large factories have been divided up into small independent work areas, and power has been delegated to lower-tier executives and to foremen. The importance of this kind of co-determination, of better information for employees and of continuous training is obvious to both employers and unions. The influence of the individual is also stressed in the so-called Development Agreement of 1982 between SAF and LO/PTK (see below).

Company participation refers to the influence of employees over their workplace through representative channels. In Sweden this is done entirely through the trade unions. At this level, union representation in advisory bodies within companies began when the SAF-LO and SAF-TCO works council agreements were signed in 1946. These agreements were replaced in 1977 by the Act on Employee Participation in Decision-Making (MBL).

Employee representation on boards of directors was introduced in 1973. Its aim was to give the unions increased information while giving boards of directors the benefit of experience accumulated by employees. The Act entitles local unions to appoint two regular members and two deputy members to the boards of virtually all companies or corporate groups with at least 25 employees. The deputies are entitled to attend meetings and speak, but have no voting right. Employee members are accorded essentially the same rights as the other board members.

MBL has attracted great attention. This law applies to all workplaces where one or more union members are employed. The law also regulates numerous other issues, such as the right of association and negotiation, and rules concerning mediation and conciliation.

The employer also has a primary obligation to provide information. Unions have access to practically all company documents. At the request of the union, employee participation in decision-making can be further regulated through a collective bargaining agreement. In certain types of dispute, the view of the unions enjoys priority until the dispute has been settled in the Labour Court or by other means. Trade unions have some veto rights against the temporary use of sub-contractors.

Supplementary employee participation agreements have subsequently been concluded in most sectors, beginning with the State civil service in 1978. SAF, LO and PTK reached such an agreement in 1982. The main concept of this Agreement on Efficiency and Participation (in Sweden called the Development Agreement) is decentralization to foremen and individual employees. All the nation-wide employee participation agreements that have been signed since MBL went into effect are supposed to be followed up by local collective agreements.

Financial participation, lastly, has come to focus on the question of who should own companies. In 1975 LO called for the creation of employee funds, sometimes also referred to as wage-earner funds. This issue has awakened by far the greatest disagreement between the employer and employee sides in recent years.

A number of models for such funds were presented. In 1983 Parliament took the decision to introduce a system of five regional funds as from January 1984. Each fund has a nine-member board of directors including five trade union representatives. Roughly SEK 2 billion per year is collected from companies in the form of 20% of net profits exceeding a certain level (after adjustment for inflation) plus an employer fee totalling 0.2% of payrolls. According to the Act this system will last for seven years from 1984 to 1990.

Employment

In international terms, Sweden has a high participation level in the labour market. The labour force of 4.4 million people covers over 52% of the overall population and nearly 82% of those aged 16–64. 1.1 million people in the labour force work part-time.

The sharp rise in the number of part-time employees has primarily been the result of the very large increase in the number of women on the job market. During the 1940s women accounted for about 25% of the labour force. By 1986 this figure had climbed to around 48%.

By law, the maximum normal working week is 40 hours. Actual working time, i.e. after taking into account absences and overtime, is about 1,513 hours per year. Internationally speaking, this is a very low figure. According to the International Labour organization (ILO), in 1984 the average working week in Sweden was 35.7 hours, compared with, for example, 44.5 hours in Switzerland.

Registered unemployment in Sweden is at around 2%. In addition, 3–4% of the labour force has been involved in government-sponsored labour market policy programs. Responsible for overseeing these programs is the National Labour Market Administration, consisting of the National Labour Market Board (*Arbetsmarknadsstyrelsen, AMS*)—whose governing body has a majority of union and employer representatives—the county labour boards and the local employment service offices.

Labour market policy focuses mainly on job placement services. During fiscal 1986/87 there were a total of 894,000 people registered at the employment services as job seekers. The total number of registered job openings was 713,000.

The most important programs for those to whom no jobs could be offered on the open market were relief work (17,600 people), labour market training for an average of 34,000 people at a time, and employment with government-subsidized pay for 41,500 people. An additional 28,000 people were employed in sheltered workshops run by *Samhall AB,* founded in 1979. Pure unemployment compensation normally only accounts for one-tenth of the costs of labour market policy, but in recent years it has climbed to more than one-fourth.

Unemployment insurance is administered mainly by unemployment benefit societies linked to various trade unions. These societies are formally independent of the unions, however, and are financed primarily by the national government.

Current topics

For many years, the Swedish labour market was able to expand on the basis of an economy that grew stronger year by year. The country could afford rising unit labour costs. In 1974 Sweden passed the United States in terms of unit labour costs and largely maintained the world's highest such costs until 1982. The devaluation of the Swedish krona by 16% in that year, however, improved the competitive position of Swedish companies and industry has been doing well since then.

At the same time, public expenditure grew to the point where it now constitutes some 70% of GDP—a very high figure in international terms. Nearly half of this outlay is made up of transfer payments within the social welfare system (pensions, child allowances, etc.)

Because of growing oil prices in the 1970s and major increases in unit labour costs in the middle of the same decade, the Swedish economy was no longer capable of bearing the costs of these large-scale public commitments. In order to combat unemployment and maintain its social welfare safety net, Sweden borrowed heavily abroad. The national government budget deficit grew rapidly from 1976.

During the 1980s, however, major efforts have been undertaken to put a halt to this dependence on loans and to keep inflation down. An important goal is to be able to continue to meet existing needs in health care, social services, education and the labour market. Many of these measures have been quite successful. Entrepreneurial activities have been much more accepted than in the 1970s.

References

Appendix 2 contains a separate list of references to the RAM project research reports (RAM reports 1–12).

Adriansson, L., Nordling, J., Sandberg, A. and Söderström, R. (1982). *A Way Towards the Office of Opportunities, a futurological study of office information systems.* (Lund: Studentlitteratur).

Bjerin, A. and Martens, P. (1981). *The Effects of the Work Environment on Leisure Time, Family and Health.* The RAM report no. 10.

Bolinder, E. and Ohlström B. (1971). *Stress at Swedish Workplaces.* A questionnaire survey among members of trade unions affiliated to the Swedish TUC (LO). (Lund: Prisma in cooperation with the LO.)

Bradley, G. (1970). *Women and Careers: A study of women's interest in promotion in relation to job satisfaction, the home environment and schooling.* (Stockholm: The Swedish Council for Personnel Administration).

Bradley, G. (1970). *Attitudes to Work and Interest in Promotion Among Female Bank Employees.* (Stockholm: The Swedish Council for Personnel Administration).

Bradley, G. (1972). *Woman and Career.* (Stockholm: The Swedish Council for Personnel Administration).

Bradley, G. (1975). Is the Computer a Workmate? *Research and Progress* **6**, 22, 25, 48.

Bradley, G. (1977). *Computer Technology, Working Life and Communication.* The Delegation for Long-Term Motivated Research (Stockholm: Liber förlag).

Bradley, G. (1979). *Computer Technology and Working Life.* A project programme for The Work Environment Fund.

Bradley, G. (1979). *Occupational Roles and Living Environments.* (Stockholm: Wahlström & Widstrand).

Bradley, G. (1981). *The Working Environment and VDTs.* Analyses of VDT Systems—Psychosocial Aspects of the Working Environment. The Work Environment Fund.

Bradley, G. (1983*a*). *Computers and their Psychosocial Impact, especially with regard to Women.* Report presented at the International Women's Seminar in Salzburg, 1982. RAM Report no. 11.

Bradley, G. (1983*b*). Effects of Computerization on Work Environment and Health From a Perspective of Equality Between Sexes. *Occupational Health Nursing Journal,* **31**, (11) 35–39.

Bradley, G., (1985). *Computers and Work Content, Work Load and Stress—Analyses and Women's Participation Strategies.* (Amsterdam: North Holland Publishing Co.)

Bradley, G., Börjeson, K. and Lundgren, M. (1973). *The AFTA formula* (Unpublished).

Bradley, G., Börjeson, K. and Lundgren, M. (1974). *The Work Environment and Salaried Employees.* (Stockholm: A Swedish Confederation of Salaried Employees report).

Bradley, G. and Nilsson, I. (1976). *Work Environment and Computerization.* Preparatory quantitative analyses of the relationship between administrative rationalization and psychosocial aspects of the work environment.

Bradley, G., Bergström, C. and Sundberg, L. (1984). *The Role of the Secretary and Word and Text Processing – Change in the Office Profession.* (Stockholm: The Work Environment Fund).

Bronner, K. and Levi, L. (1967). *Stress in Working Life.* (Stockholm: The Swedish Council for Personnel Administration and The Swedish Industrial Salaried Employees' Association).

Cronbach, L. J. (1951). Coefficient of Alpha and Internal Structure of Test. *Psychometrica* **16**, 297–334.

Dahlström, E. *et al.* (1966). *Technical Change and Work Adjustment.* (Stockholm: Prisma).

Docherty, P. *et al.* (1977). *How to Succeed in Systems Development.* (Stockholm: EFI).

Docherty, P. and Loman, L. (1983). Personnel Influence and Computerization. (Stockholm: PTK and EFI).

Ehn, P. and Sandberg, A. (1979). *Company Management and Wage-Earner Power.* (Stockholm: Prisma in cooperation with The Swedish Center for Working Life).

Frankenhaeuser, M. (1974). *Stress, Tolerance Limits and The Quality of Life.* Report no. 1. (Stockholm: The Institute of Psychology).

Frankenhaeuser, M. and Ödman, M. (1983). Stress—a Part of Life. (Stockholm:Brombergs Bokförlag).

Gardell, B. (1979). *The Work Environments of Salaried Employees. The Psychosocial Work Environment and Health—A Preliminary Report.* (Report no. 24.) (The University of Stockholm: The Institute of Psychology).

Grip, A. and Sundström, L. (1984). *The Postal Giro Centre in Stockholm.* (Stockholm: The Swedish Center for Working Life).

Goldthorpe *et al.* (1971). *The Affluent Worker: Industrial Attitudes and Behaviour.* (Cambridge: Cambridge University Press).

Göransson, B. *et al.* (1974). *The ALLFA Survey: Case Studies in the PAAS Project, a Work Report.* (Stockholm: The Swedish Center for Working Life. 1979: 14).

Gunnarsson, E. and Söderberg, I. (1979). *Work on Text Display Units in Newspaper Companies: a Sight-Ergonomic Inventory.* (The National Board of Occupational Safety and Health, Research Report 1979:21).

Ingelstam, L. (1981). *The Value of Work and the Use of Time.* (Stockholm: Liber Förlag—The Secretariat for Futurological Studies. Source Documents 12–16, The Coordinating Board of the Swedish Research Councils).

Johansson, G. and Aronsson, G. (1984). Stress Reactions in Computerized Administrative Work. *Journal of Occupational Behaviour*, **5**, 159–181.

Jonasson, S. (1985). *Computers and the Swedish Society.* (IFIP Conference. Stockholm: Third Conference on Human Choice and Computers).

Junberger, L. (1983). *Teledata for Sweden.* (Stockholm: Liber Förlag).

Kristiansson, L. (1982). *Computers, Models, Reality.* (Stockholm: Kristiansson and Förlagsgruppen).

Levi, L. (1972). Stress and Distress in Response to Psychological Stimuli, Dissertation at Karolinska Institutet. (Stockholm: Norstedt and Söner) *Wiksell.*

Lennerlöf, L. (1981). *Operations Management Under Change – a Psychosocial Perspective.* (Stockholm:Liber)

Liljeström, R. (1973). *Social Background: The Interaction between adults and Children in a Changing Society.* (Stockholm: Allmänna Förlaget/ Publica).

Modig, M. and Sjögren, J. (ed.) (1982). *Leaders and the Led.* (Stockholm: Norstedts).

Palme, J. (1975). *People in the Computer Society.* (Lund: Studentlitteratur).

Persson, S. and Strandler, L. (1978). *Decentralization and Computers.* (Stockholm: The Swedish Confederation of Employers (SAF)).

Sandberg, Å. (1984). *Between Alternative Production and Industrial Research and Development*—Report from the Utopia Project, (Stockholm: The Swedish Center for Working Life).

SACO/SR (The Swedish Confederation of Professional Associations) *SACO/SR and Computerization.* A debate paper submitted to the 1982 congress.

The SAF-PTK Rationalization Council. (1974). *The New Rationalization.*

SOU 1984:20 (The Swedish Government Official Reports). *Computers and Change in Working Life.* The Final Report of the Committee on the Effects of Data Processing Technology.

SOU 1984:51 (The Swedish Government Official Reports). *Data Processing Technology and Industrial Renewal.* The Final Report of the Committee on Data and Electronic Technology.

Söderström, M. (1983). *Educational Research in Working Life.* Report, 1982:19. (Stockholm: The Swedish National Board of Universities and Colleges, UHÄ).

The Swedish Federation of Data Processing Users. (1980). The 1980s on a card. Data Technology: Development and Environment 1980–1990.

The Swedish Confederation of Employers. (1972). *The Users Develop-ADP.*

The Swedish Confederation of Employers. (1976). *AR- A Way of Developing.*

The Swedish Confederation of Salaried Employees (TCO). (1984). *The Right Way of Working on VDUs.*

The Swedish Confederation of Salaried Employees (TCO). (1985). *Creativity in Working Life—on Professional Knowledge, Work Organization and New Technology.*

Swedish Government Bill. (1981/2). 1981/1982:123. *Coordinated Data Policy.*

Swedish Government Bill. (1984/5). 1984/85:220. *Data Policy.*

Swedish Government Bill. (1985). 1985/85:225. *EDP in Government Administration.*

The Swedish National Board of Occupational Safety and Health. (1980). *Psychological and Social Aspects of the Work Environment.* (1980:14).

The Swedish National Board of Occupational Safety and Health. (1982). *The Code on Solitary Work.* (1982:3).

The Swedish National Board of Occupational Safety and Health. (1983). *The Code on Work Positions and Work Movements.* (1983:6)

The Swedish National Board of Occupational Safety and Health. (1985) *Work on VDUs.* (1985:12).

The Swedish National Board of Occupational Safety and Health for Government Employees. (1978). *Terminal Work, Micro Picture Reading.* (Stockholm: Liber).

The Swedish National Union of Insurance Company Employees. (1975–85). *RAM Reports nos. 1–12. Guidelines and Directions for People Responsible for Data.*

The Swedish National Union of Insurance Company Employees. (1978) *Co-determination in Advanced Technology.*

Toffler, A. (1970). *Future Shock* (London: Pan Books)

UHÄ. (1982). (The Swedish National Board of Universities and Colleges). *Personnel and Working Life Issues.* Report 1982:19

Unga, N. (1980). *The New Technology and Our Jobs—Facts and Personal Reflections*. Extract from a TCO debate paper at the Conference 'The Development of Computer Technology—Control or Be Controlled?'

Vroom, V. (1964). *Work and Motivation*. (New York: John Wiley & Sons).

Westin, B. (1975). *Work at VDTs*. no. 1. (Stockholm: Arbetsmilgö).

Westlander, G. (1978). *What are Psychosocial Issues?* (Stockholm: The Joint Industrial Safety Council—The Work Environment Fund).

Index